Bell Employees – Wheatfield Plant

Other Books by August A. Cenkner Jr.

- A Dark Energy Theory Correlated With Laboratory Simulations and Astronomical Observations, ISBN: 1-4208-3447-9 (sc), 08/23/2005.

- Dark Energy – Laboratory Simulations Lead to Predictions of: Star Accelerations; Formation of Spiral Galaxies; Creation of Voids, Walls, and Clusters, ISBN: 978-1-4343-0661 (sc), 08/22/2007.

- Hubble Space Telescope Identifies Dark Energy, ISBN: 978-1-4490-1134-5 (sc), 08/03/2009.

Aerospace Technologies

of

Bell Aircraft Company

A Pictorial History

1935-1985

August A. Cenkner Jr.

AuthorHouse™
1663 Liberty Drive
Bloomington, IN 47403
www.authorhouse.com
Phone: 1-800-839-8640

First published by AuthorHouse   05/13/2011

ISBN: 978-1-4634-0213-6 (sc)
ISBN: 978-1-4634-0453-6 (ebk)

Library of Congress Control Number: 2011908088

Printed in the United States of America

Any people depicted in stock imagery provided by Thinkstock are models, and such images are being used for illustrative purposes only.
Certain stock imagery © Thinkstock.

This book is printed on acid-free paper.

Because of the dynamic nature of the Internet, any web addresses or links contained in this book may have changed since publication and may no longer be valid. The views expressed in this work are solely those of the author and do not necessarily reflect the views of the publisher, and the publisher hereby disclaims any responsibility for them.

To Judy

my wife and best friend

in celebration of being together for 50 wonderful years

ACKNOWLEDGEMENT

The author would like to express his appreciation to the staff and volunteers of the Ira G. Ross Aerospace Museum in Buffalo, New York, for allowing me unlimited access and unlimited use of the Bell-aerospace-company's historical information.

The author selected all of the material that is included in this book. The book relies exclusively on hardcopy documentation. This implies that if a certain technology was not available in The Museum archives, then it is not addressed in this book. Unless otherwise acknowledged, all works of art – photographs, drawings, flow charts, graphs, tables, slides, and Bell product information – were obtained from The Museum. Frequently, for clarification, the author added descriptive information to Bell's art work. Unfortunately, much of the original art work was not available, so the use of reproductions became the rule instead of the exception. Frequently, this art work was not dated or annotated and personnel were not identified.

The main thrust of this book is to acknowledge the technologies that the Bell-aerospace-company developed or refined. If certain programs incorporated technologies that were basically the same as other programs, then these same technology programs were not included in detail.

The author provided his own funding, and all other resources, that were required to research, design, write, type, edit, and publish this book.

On behalf of all the people of Western New York, and all the Bell employees who no longer live in the area, the author would like to thank the many volunteers who helped create and maintain the Ira G. Ross Aerospace Museum. The Museum could not exist without their help and dedication, and our spectacular aerospace history would have been lost and unknown to future generations,   THANK YOU !

Many of the technologies that are discussed in this book can be viewed at the Ira G. Ross Aerospace Museum in Buffalo, New York. For further information, visit www.wnyaerospace.org.

Lawrence Dale Bell and Associates

x

Lawrence Dale Bell

April 5, 1894 – October 20, 1956

U. S. A. President Franklin D. Roosevelt meets with Larry Bell and visits the Bell Aircraft Company.

Larry Bell shares 1947 Collier Trophy with pilot Chuck Yeager and research scientist John Stack.

Lawrence Dale Bell Presentation at U.S.A.F. Academy, Colorado. November 12, 1969.

The 10,000th P-39, manufactured by Bell Aircraft Company, is introduced by Larry Bell.

xii

Members of Bell employee War Years Inventors Group, 1941.

Speakers table at 1952 Management Dinner. Maj. Gen. Olmsted, Adm. Ramsey, Larry Bell, L Faneuf, Lt. Gen. Craigle.

10 year pins are awarded to Bell employees by Larry Bell, 1948.

Veterans of Foreign Wars hang flag in Bell Aircraft office; Larry and Vaughn Bell present.

Table of Contents

1.0 History of the Bell-aerospace-company

2.0 Air Craft

List of Tables

Introduction

When Lawrence D. Bell formed the original Bell-aerospace-company -- the Bell Aircraft Company -- on July 10, 1935, his goal was to develop and manufacture various types of airplanes. He had no way of knowing that he had unleashed a technological phenomena that would extend its influence beyond fixed wing airplanes and into the realm of many other types of air craft, as well as various types of space craft, water craft, and land craft; many of these technologies would then be transferred into other areas to address unrelated problems.

The original 50 employee company would grow to more than 50,000 and the company would expand throughout the United Stated and into Canada. While the original Bell company in Buffalo New York no longer exists, many of the developed technologies, as well as numerous spun-off companies, have survived to this day. However the original Wheatfield plant/Bell headquarters, Figure 1.6-d, still exists; sections of the complex have been rented and are used by numerous small companies.

This book introduces the technologies that the Bell-aerospace-company (or Bell) (i.e. first the Bell Aircraft Company and then the Bell Aerospace Corporation and finally Bell Aerospace Textron (Section 1.1)) created, developed, extended, manufactured, and/or proposed.

All things considered, it seemed logical to subdivide the book into five major sections:

(1) History of the Bell-aerospace-company -- to discuss how the company grew and evolved over the years and to give a historical perspective on when Bell was involved in the various technologies.

(2) Air Craft – how Bell was involved in developing many types of craft that flew in the air, including peripheral support equipment.

(3) Space Craft -- Bell's involvement in developing craft and support equipment that was used in space to orbit the earth or explore the moon.

(4) Water/Land Craft – contributions, made by Bell, to the development of crafts and support equipment that traveled over water and/or land.

(5) Technology Transfer – technologies that were developed for use in the above areas that were transferred, by Bell, into unrelated areas to solve new problems.

1.0 History of the Bell-aerospace-company

The original company, the Bell Aircraft Company, was first formed by Lawrence D. Bell on July 10, 1935. As it grew and expanded, to accommodate the demands of

WW II, various divisions were formed and its name was changed to the Bell Aircraft Corporation. Around June, 1960 the company was purchased by Textron Inc. Its name was changed to the Bell Aerospace Corporation and it was subdivided into three companies: the Bell Aerosystems Company, the Bell Helicopter Company, and the Hydraulic Research and Manufacturing Company.

A reorganization occurred in January 1970 followed by a name change to Bell Aerospace Textron. In April 1998, the original Bell Company ceased to exist when the Wheatfield Plant was sold to the Industrial Realty Group. However many of the developed technologies, and spun-off companies, survived and continue to exist to this day; see Section 1.1.

Section 1.2 summarizes the main missions of the various divisions and companies of the Bell-aerospace-companies.

A detailed listing of the capabilities of one of the Bell-aerospace-companies – the Bell Aerosystems Company – appears in Section 1.3 -- which is followed by a technology development time line in Section 1.4.

Bell used discretionary funds to support in-house Individual Research and Development Projects. Projects that were selected for this funding during two randomly selected sample years -- 1967, and 1970 – are presented in Section 1.5.

Location of the three Bell Aerosystems Company facilities in Western New York are shown in Figure 1.6-a. Details on the Air Cushion Vehicle Base , the Wheatfield Plant, and the Bell Test Center are given, respectively, in Figures 1.6-c, 1.6-e, 1.6-f.

2.0 Air Craft

Section 2.1.1 lists the fixed wing airplanes that were developed, manufactured, and/or proposed by Bell; they are not listed in any particular order. There was a gradual transition from propeller driven craft to jet and rocket craft with an accompanying increase in ceiling and maximum level speed.

To put things into perspective time-wise, Figure 2.1.2 highlights the development of various fixed wing and helicopter airplanes. Significant technological achievement with airplane performance, and with other technologies, is highlighted in Section 2.1.3.

Sections 2.1.4 to 2.1.19 briefly summarize information about the airplanes that Bell was involved with.

A small commercial LA-5 airplane and a large military XC-8A military cargo plane were retrofitted with air cushion landing gear, Section 2.2, and successfully tested. The planes could take off and land on runways that were not accessible to planes with conventional wheeled landing gear.

Extension of rocket/jet back pack and platform flying was first extended to vertical/short take-off and landing airplanes with the bell "air test vehicle"; Section 2.2. The basic concept was then extended to the X-14 and the D-188A jet airplanes.

Bell demonstrated the V/STOL technology with the XV-15 and the XV-3 propeller driven airplanes; Section 2.3.4 and 2.3.5.

To increase the V/STOL payload a four engine propeller craft was developed, known as the X-22A; Section 2.3.6. The X-22A was actually the last military airplane built in Western New York.

In Section 2.4.1 an abbreviated history of the development of helicopter technology, by the Bell Aircraft Corporation, is outlined..Starting with the small scale testing by Arthur Young, it is traced through to modern helicopters.

Personal "flying" machines are the topic for Section 2.5. It begins with a tabular listing of six different rocket or jet propelled flying machines for one or two people. Detail is then given on each of these machines.

Bell was heavily involved in the development of military missiles, as summarized in Table 2.6-a, for both domestic and foreign use. Its' contribution spanned the testing of a foreign missile, in addition to the development and manufacture of missile components and complete missiles.

To improve missile efficiency, and therefore increase range or payload, Bell invented an extendable nozzle cone for it's rocket engines; Section 2.6.8.

From Section 2.7, Bell even got involved, briefly, in the manufacture of a guided bomb.

Bell developed and deployed eight different microwave landing systems, Section 2.8.2, for remotely controlled landing of airplanes on aircraft carriers in any type of weather, a fixed land based version of the same system, a small mobile land based version of the system, and a fixed land based version for the Air Force.

Various types of airplane communication antennas were developed and deployed in the field, Section 2.9, for military use.

According to Section 2.10, a unique rotating arm hypersonic test facility was built.

to evaluate/develop different types of material. Of particular concern were airplane locations where materials were exposed to severe weather related erosion problems, while flying at hypersonic speeds.

In the area of electronics, Bell developed and produced inertial navigation systems, accelerometers, gravity meters, and gyroscopes; Section 2.11.

Extensive computer programming support, Section 2.12, was required to operate the various types of computer controlled equipment supplied by Bell.

3.0 Space Craft

In the Space program, Bell made significant contributions in the areas of : liquid storage and expulsion tanks, rocket engines for second stage launch vehicles, reaction control rocket engine systems for controlling vehicle attitudes, Apollo lunar lander ascent engine, astronaut training of simulated landings on the moon, and Gemini docking of space vehicles.

As detailed in Section 3.1, positive expulsion liquid storage tanks were used extensively on Apollo mission vehicles.

In addition to building the primary engine, the secondary propulsion system for the Agena Target Vehicle and the lunar module ascent engine, five generations of Bell's Agena rocket engine were built for use on upper stages of various launch vehicles; Section 3.2.1 and 3.2.2.

With various types of rocket engines, Bells' reaction control systems, Section 3.2.3, were employed to control the attitude of various types of vehicles. Section 3.2.4 reviews the development of more efficient rocket engine propellants.

In preparation for landing on the moon during the Apollo mission, Bell built training vehicles for astronauts to practice moon landings; Section 3.3. Concepts for flying vehicles to explore the moon; Sections 3.3.3 and 3.3.4, were never realized.

Proposals for rocket launched space-glider-planes, Section 3.4, were never pursued in the time frame covered here.

Software and hardware were created, Section 3.5, for astronauts to practice simulated docking of the Gemini and Gemini-Target-Vehicle and for landing on the moon during the Apollo mission.

Satellite tracking and communication systems were also part of Bell's electronics contribution; Section 3.6.

In Section 3.7, a specially built space simulation laboratory is introduced. It was utilized to test a remotely controlled robot that was to be used for various types of space missions, like remote spacecraft repair, personnel transfer between vehicles and resupply.

Section 3.8 details Bell's contribution to the successful space docking of the Gemini spacecraft and the Gemini-target-vehicle, in preparation for the Apollo mission.

The space craft section finishes with a proposal to utilize Bell's expertise in the creation of the space shuttle.

4.0 Water/Land Craft

The most significant contribution that Bell made, in the area of water/land craft, is the development of air cushion vehicle (ACV) technology. Next would be surface effect ships, and finally military armored tanks with weapon systems.

Section 4.1 details the history on the evolution of Bells' air cushion vehicles and surface effect ships.

A schematic of a typical air cushion vehicle is shown in Section 4.2.

The development of ACV technology required the design and construction of a specialized sophisticated small scale test facility. Details on this facility are shown in Section 4.3.

Sections 4.4 to 4.10 present the capabilities of various types of ACV's developed by Bell, while Sections 4.11 to 4.13 shows some surface effect ships.

Proposals for the development/manufacture of military combat tanks are detailed in Sections 4.14 to 4.16.

5.0 Technology Transfer

Technologies and experience gained, while working in the above areas, was transferred to new areas to help solve unrelated problems.

Gravity meter technology was transferred to the exploration of oil, in Sections 5.1.1 and 5.1.2.

Military ground communications in forward areas was enhanced when a lightweight troposcatter communications set was designed; Section 5.2.

Rocket engine combustion technology was employed in the development, in Section 5.3, of a system to convert coal into natural gas.

Rocket engine combustion technology was again employed to the development of chemical lasers; Section 5.4.

Technology created while Bell was engaged in jet/rocket back pack work was transferred to the creation of a special increased-load back pack, in Section 5.5.

In Section 5.6, experimental work that was acquired while developing high energy gas dynamic lasers, was transferred to the astronomical area to gain insight into what dark energy is and to explain Hubble Space Telescope observations.

In an attempt to acquire non-military work, at a time when there was a major cut-back in military contracts, Larry Bell -- in Section 5.7 -- invented and manufactured the prime mover using the expertise acquired earlier on aerospace projects.

Technology Trends

It's a fascinating journey to follow the logical expansion of Bell's technologies into other areas. For example, fixed wing propeller driven airplanes lead to jet powered and rocket powered airplanes, which spun off to: personal air craft that were jet and rocket engine powered back packs, rocket powered vehicles that were ridden by the pilot, a rocket powered personal space back pack and craft for use in space, a rocket powered craft for exploring the surface of the moon, and the jet/rocket powered Lunar Lander Simulator, for the Apollo mission, to practice landing on the moon. This background leads naturally to developing/manufacturing rocket engines for missiles and space craft launch vehicles, in addition to rocket engine systems for controlling the attitude of various types of craft. This rocket engine experience was then transferred to the development of coal gasification plants, chemical lasers, and the identification of dark energy.

In stark contrast, Bell had the insight to expand, from fixed wing airplanes into rotary wing airplanes, when a young inventor, Arthur Young, demonstrated a working model of the first known flying helicopter. Other companies had seen the model but they had all rejected the technology. This was a major victory for the company. It eventually led to the creation of a company dedicated to helicopters.

Bell's involvement in the electronics area appears to have been prompted, in large part, by the need to develop radio controlled model air cushion vehicles, a remotely controlled Grumman F7F Tigercat air plane and a command control system for a surveillance drone. In this area, a major accomplishment was the development and deployment of numerous automated microwave landing systems for the Navy and the Air Force. In their aerospace business, this expertise was also utilized in developing remotely controlled space robots.

When developing new technologies, the Bell-aerospace-company frequently, and effectively, relied on small scale simulations to guide this development. Consider for example:

(i) Airplane models used to evaluate different designs in wind tunnel testing. Consider the P-39 in Figure 2.1.5-i.

(ii) Helicopter (rotating wing airplane) models developed to solve the vibration problem. Figure 2.4.2-a.

(iii) Shrike guided missile – a reduced size version of the RASCAL air-to-surface missile. Figure 2.6.6-a.

(iv) Material evaluation/development, for hypersonic flight, using a specially built rotating arm laboratory. Figure 2.10-b.

(v) Computer controlled simulator for astronauts to practice the space rendezvous and mating of the Gemini and Gemini Target spacecraft. Figure 3.5.4-a.

(vi) Computer controlled simulator for astronauts to practice flying over the moon's surface and then landing. Figure 3.5.7-c.

(vii) Simulation of a remotely controlled space robot in a simulated space environment. Section 3.7.1.

(viii) Small scale models of air cushion vehicles, tested in a specially built water tank laboratory. Figures 4.3-a and 4.3-b.

(ix) A special supersonic wind tunnel to study and solve the laser cavity reactant mixing problem. Figure 5.4.3-a.

(x) A full size wooden mockup of a lunar flying vehicle, built for a standing astronaut as demonstrated in Figure 3.3.3-b.

A major consideration, in selecting the type of book to create, was the type of information that was available in The Museum. Because of the time periods involved, it seemed advisable to rely exclusively on written documentation instead of the memories of past employees, who were still alive. Cited written documentation was obtained almost exclusively from The Museum.

To meet these objectives it seemed that a large format pictorial color book, with a landscape format, was best suited since large photographs, capable of showing more detail, could be utilized. However it seemed advisable to minimize the selling price of the book, so color was eliminated.

Book Philosophy

While the primary objective of this book is to document technological accomplishments of the Bell-aerospace-company, it became clear that it should also strive to serve a number of other objectives:

(1) Preserve as much technological detail as possible so future generations can appreciate how these technologies worked.

(2) Attempt to stimulate the scientific interest of younger generations.

(3) Educate interested parties on the strategies that were used to develop these, at the time, ground breaking technologies.

(4) Demonstrate the usefulness of employing simulations in the development of new technologies.

(5) Point out the natural progression in the evolution of more advanced technologies, from earlier technologies.

(6) Where possible, identify references where more detailed technology information can be found.

1.0   History of the Bell-aerospace-company

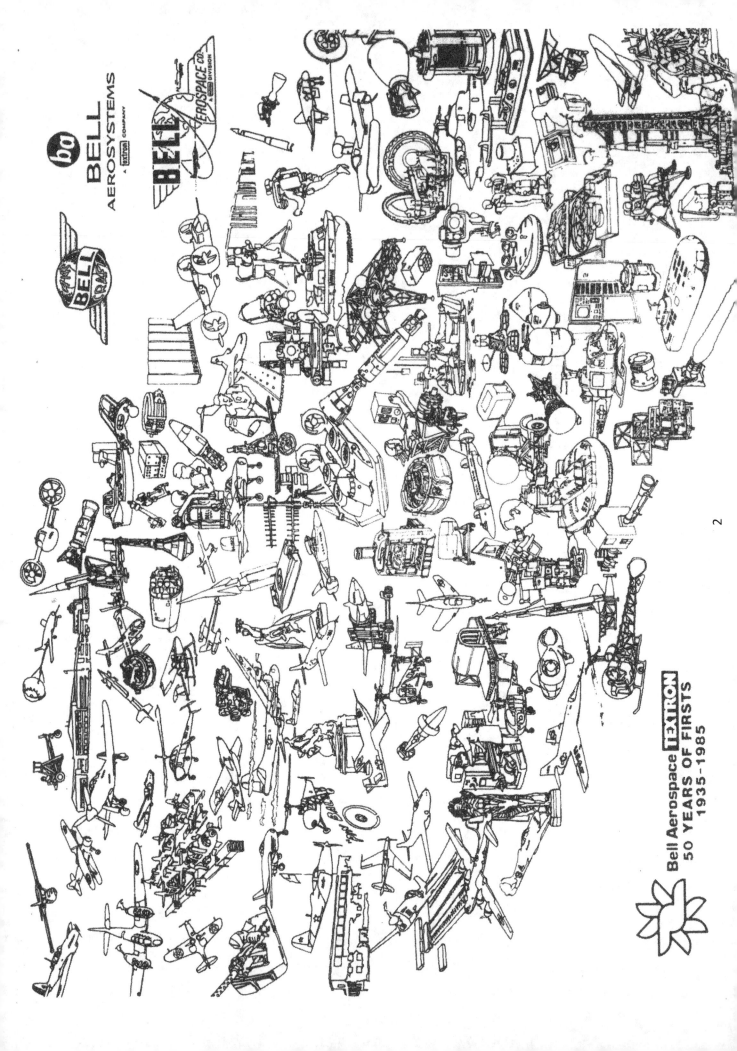

Bell Aerospace **TEXTRON**
50 YEARS OF FIRSTS
1935-1985

2

1.1 Organizational History of the Bell-aerospace-company

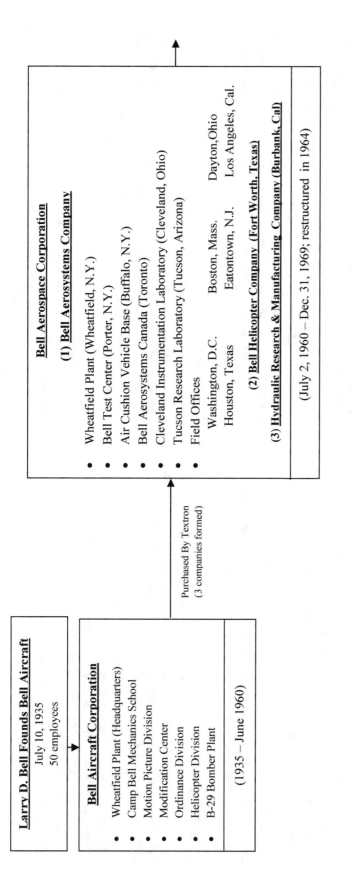

Larry D. Bell Founds Bell Aircraft

July 10, 1935
50 employees

Bell Aircraft Corporation

- Wheatfield Plant (Headquarters)
- Camp Bell Mechanics School
- Motion Picture Division
- Modification Center
- Ordnance Division
- Helicopter Division
- B-29 Bomber Plant

(1935 – June 1960)

Purchased By Textron
(3 companies formed)

Bell Aerospace Corporation

(1) Bell Aerosystems Company

- Wheatfield Plant (Wheatfield, N.Y.)
- Bell Test Center (Porter, N.Y.)
- Air Cushion Vehicle Base (Buffalo, N.Y.)
- Bell Aerosystems Canada (Toronto)
- Cleveland Instrumentation Laboratory (Cleveland, Ohio)
- Tucson Research Laboratory (Tucson, Arizona)
- Field Offices
  - Washington, D.C.   Boston, Mass.   Dayton, Ohio
  - Houston, Texas   Eatontown, N.J.   Los Angeles, Cal.

(2) Bell Helicopter Company (Fort Worth, Texas)

(3) Hydraulic Research & Manufacturing Company (Burbank, Cal)

(July 2, 1960 – Dec. 31, 1969; restructured in 1964)

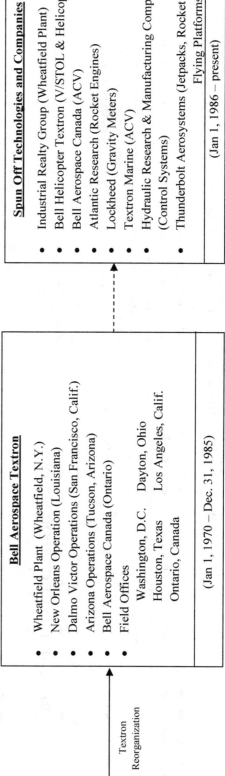

Bell Aerospace Textron

- Wheatfield Plant (Wheatfield, N.Y.)
- New Orleans Operation (Louisiana)
- Dalmo Victor Operations (San Francisco, Calif.)
- Arizona Operations (Tucson, Arizona)
- Bell Aerospace Canada (Ontario)
- Field Offices
  - Washington, D.C.   Dayton, Ohio
  - Houston, Texas   Los Angeles, Calif.
  - Ontario, Canada

(Jan 1, 1970 – Dec. 31, 1985)

Textron
Reorganization

Spun Off Technologies and Companies

- Industrial Realty Group (Wheatfield Plant)
- Bell Helicopter Textron (V/STOL & Helicopter)
- Bell Aerospace Canada (ACV)
- Atlantic Research (Rocket Engines)
- Lockheed (Gravity Meters)
- Textron Marine (ACV)
- Hydraulic Research & Manufacturing Company (Control Systems)
- Thunderbolt Aerosystems (Jetpacks, Rocket belts, Flying Platforms)

(Jan 1, 1986 – present)

Cenkner Flow Chart

3

1.2   Discussion of the History of the Bell-aerospace-company

The Bell-aerospace-company, over the years, has experienced numerous name changes, spectacular growth accompanied by reorganizations into new divisions and new companies, realignments, the sale of the company, and phenomenal technological achievements; see Figure 1-1.

It all began on July 10, 1935 when Lawrence D. Bell founded the Bell Aircraft Company in Buffalo New York by taking over vacated facilities, and experienced past employees, of Consolidated Aircraft; initially, there were 50 employees.

BELL AIRCRAFT CORPORATION
(July 10, 1935-June 1960)

Wheatfield Plant (Headquarters)
Develop, manufacture, and test various types of aircraft, spacecraft, and watercraft. Time and labor saving processes, operations, devices, and machines – by the hundreds – were created. Evolved and extended new technologies including the technology transfer to new areas.

Camp Bell Mechanics School
Bell set up separate training facilities, near the production plant, to train thousands of military personnel in the construction and maintenance of the P-39.

Motion Picture Division
This division was very active in creating primarily military motion pictures, in black/white or color, and sound. It even created some animated movies. Through 1945, close to 100 projects had been recorded, that required 3,500,000 feet of 16 mm film stored in 125 reels:

The Airacobra	Cannon on Wings	Introduction to the P-39
Flying the P-39	Bell Helicopter	B-29's Over Dixie
P-63 Pilot Training	P-63 Service Operations	RP-63 Armored Airplane
It's Your War Too	Meet Your Neighbor	Important Company Events
Report on Jet Propulsion		

Construction and Flight Testing of P-39, P-63, XP-77, XP-59A, Bell Helicopter

Bell Modification Center (No. 7)
A separate factory was set up, with 925 employees, which took planes coming off the production line and made any required modifications. This included: (1) winter-ization of P-39 Airocobras that were to be used in the Arctic regions, (2) retrofitting planes for reconnaissance and photographic missions, (3) alter P-39's for special dive bombing and skip bombing missions, and (4) upgrades to the basic P-39 design. Modifications were also made on various other types of aircraft, which were built by other companies.

Ordinance Division
The recoil of .50 caliber guns, as well as other types, poses a threat to the airplane structure and gun accuracy. Bell developed an energy absorbing mount that mitigated this threat. These gun mounts, totaling more than 375,000, were eventually installed in 19 different airplanes and four different surface craft. The division was also involved in numerous other production projects including the production of 4.2 mortar shells. During its first four years, its production encompassed 104 different types of gun mounts, spare parts, ammunition boxes and related mechanisms. Additional space was required due to the rapid increase in production work. The plant, initially located in Buffalo N.Y., was relocated and expanded in Burlington, Vermont.

Helicopter Division
The original Helicopter plant was opened in Gardenville, N. Y. in an old garage, with about 15 employees. Arthur M. Young was hired to construct several full scale helicopter prototypes, Bell Models 30 and 47 that were based on his successful small scale model. Floyd Carlson, an experienced airplane pilot but self taught helicopter pilot, demonstrated the flight characteristics of the prototype Young helicopter. As the business took off, the production plant was relocated to Texas.

B-29 Bomber Plant
Bell Aircraft Corporation was selected, from among other manufactures, as the prime contractor to produce the B-29 Superfortress bomber, employing the Boeing Aircraft design. The Army Air Force selected a site adjacent to the Rickenbacker Airport at Marietta, Georgia, to build a new plant. The plant was so large that 20 of the largest battleships would fit in, along with 69 submarines, and 24 PT boats. The

production process was modified so the B-29s left the plant combat ready, after the usual flight tests; 663 B-29's were built as part of the war effort.

BELL AEROSPACE CORPORATION
(July 2, 1960 – Dec. 31, 1969)

(When Textron Inc. purchased Bell, three companies were formed.)

(1) BELL AEROSYSTEMS COMPANY

Wheatfield Plant (Headquarters)

Bell Aerosystems Company headquarters is located in the Wheatfield Plant, in the Town of Wheatfield in New York State It houses marketing, engineering, planning, manufacturing, finance, and administration departments in addition to laboratories and computer facilities. A detailed layout of this plant is shown in Figure 1-1.

Instead of remaining organized along product lines, President William G. Gisel, in the summer of 1964, reorganized Bell along functional lines: (1) Propulsion Systems and Components, (2) Aerospace Systems, (3) Electronic Systems, (4) Air Cushion Vehicles, and (5) Advanced Technologies.

Bell Test Center

Located in the Township of Porter, N. Y., it was initially used specifically for rocket engine research and development; it was then expanded for production testing. The Test Center was operated by Bell, for the USAF, under contract to the USAF.

Air Cushion Vehicle Base

Bell air cushion vehicles are tested at a separate six acre facility on the Buffalo, N.Y. waterfront. An 8000 sq. ft. main structure is utilized for vehicle storage and repair. Vehicles can be driven, a short distance, from this main structure, to a ramp that provides access to the Lake Erie.

Bell Aerosystems Canada

Manufacture and test of air cushion vehicles.

Cleveland Instrumentation Laboratory

Manufactures and tests accelerometers, acceleration sensing systems with associated analog and digital electronics and power supplies, velocity meters, and capacitive transducers – for deployment to military and commercial missiles, spacecraft, and inertial navigators.

Tucson Research

Field testing, and testing in controlled electromagnetic laboratories, of electronic and electro-optical guidance and communications systems and sub-systems.

(2) BELL HELICOPTER COMPANY

Manufacture and test of helicopters and V/STOL aircraft.

(3) HYDRAULIC RESEARCH AND MANUFACTURING COMPANY

Development, manufacture and test of control systems for aircraft and spacecraft.

BELL AEROSPACE TEXTRON
(Jan. 1, 1970 – Dec. 31, 1985)

Wheatfield Plant

Main plant and headquarters of Bell Aerospace Textron.

New Orleans Operation

Dedicated to the manufacture and testing of Air Cushion Vehicles and Surface Effect Ships. Testing takes place in Lake Pontchartrain and the Gulf of Mexico.

Dalmo Victor Operations

Develops and manufactures electro-magnetic defense systems, aerospace antennas, and electro-optical equipment.

Arizona Operations

Involved in the testing of complex electronic and electro-optical guidance and communication systems and subsystems.

Bell Aerospace Canada

Dedicated to the production of air cushion vehicles -- the 45-ton Voyageur and the 17-ton Viking -- that can also be used in the remote areas of Canada, Alaska, and the Arctic. The use of ACV's for the break-up of ice, that formed on bodies of water, was also perfected.

References

*AN AVIATION STORY : THE HISTORY OF BELL AIRCRAFT CORPORATION, Tenth Anniversary Edition, Bell Aircraft Corporation, July, 1945

* Kreiner, C. F., Director, Public Relations Department of Textron's Bell Aerospace Division, History of Textron's Bell Aerospace Division, Rendezvous, Vol. XIV/No.2/July 1975

*Norton, D. J., LARRY : A Biography of Lawrence D. Bell, Nelson Hall Inc.,1981.

*Matthews, B., COBRA! BELL AIRCRAFT CORPORATION 1934-1946, Schiffer Military/Aviation History, 1996.

* Facilities, Bell Aerosystems Company, ca. 1960

*FOCUS ... ON TEXTRON BELL AEROSPACE TODAY, Bell Aerospace Textron, ca. 1970.

1.3 Products and Services of the Bell Aerosystems Company

AIRCRAFT, MISSILES, GROUND EFFECT MACHINE SYSTEMS

VERTICAL FLIGHT SYSTEMS - Exceptional back ground in V/STOL jet fighter/bomber and ducted propeller transport development.

GUIDED MISSILES - First complete weapon system contractor responsible for management, design and production of air/ground systems.

TARGET MISSILE SYSTEMS - Fifteen years experience in design, development and production of target missile systems.

AIR LAUNCH SYSTEMS - System design and fabrication.

HEAT PROTECTION – Double-wall construction, successfully tested as answer to re-entry heating problem. Use of refractory materials for re-entry.

GROUND SERVICING EQUIPMENT - Design and fabrication of complete GSE for aircraft, missiles, and rocket engines.

GROUND EFFECT MACHINES - Winning contractor for largest U.S. ACV - Navy's 22 1/2-ton Hydroskimmer.

AIRCRAFT DESIGN - From first American jet airplane through "X" series and proven V/STOL concepts.

STRUCTURES - Lightweight heat protection and compact design.

SPACE SYSTEMS

RECOVERABLE SPACE VEHICLES - Design, test and fabrication of manned and un-manned space vehicles for controlled landings on earth or moon.

EXTRATERRESTRIAL WORKERS - Development, fabrication and evaluation of equipment for extra-vehicular manned operations in a space or lunar environment.

SPACE VEHICLES - Design, fabrication and test of satellites including deployment, maneuvering and rendezvous.

UPPER STAGES - Design, fabrication and test of space stages involving integration of structure, tankage and propulsion system.

SIMULATORS - Fixed base simulation of manned space systems for evaluation and training.

ROCKET OPERATIONS

LIQUID ROCKET PROPULSION - Rocket engines and controls, propellant tanks, positive expulsion devices, turbine pumps and pressurization systems.

HIGH ENERGY SOLID PROPELLANTS - Synthesis of new compounds for solid propellant propulsion and energy.

ADVANCED ROCKET PROPULSION - Research and development in new propellant combinations, pressurization concepts, thrust chambers, high combustion temperatures, and materials inc luding fluorine-oxidized propulsion system technology.

REACTION CONTROLS - Low-thrust propulsion systems providing vernier velocity adjustment, propellant settling and attitude orientation.

PROPULSION SYSTEM GROUND HANDLING EQUIPMENT Low thrust propulsion systems providing vernier velocity adjustment, propellant settling, attitude orientation.

PROPULSION SYSTEM GROUND HANDLING EQUIPMENT - Designed and fabri-cated to provide check-out, functional test, and servicing of propulsion systems.

CRYOGENIC PUMPS – Fifteen years experience in pump design and development for liquid nitrogen, helium, oxygen, hydrogen and fluorine.

ENVIRONMENTAL TESTING OF PROPULSION SYSTEMS - Facilities for system and component testing at simulated altitude, pressure and temperature conditions from sea level to 10-8 Torr and cryogenic to +20,OOO°F.

SMALL ROCKET LIFT DEVICE - A new dimension in mobility, the optimized rocket belt is a complete one man personal propulsion system.

ADVANCED RESEARCH

PROPULSION AND POWER:

(1) Chemical Propellants - Study and selection of new and promising propellants and fuel blends for high energy liquid propellant rocket engines. Performance cal-culations using new computer programs for evaluating performance characteris-tics of propellant and oxidizer combinations.

(2) Nuclear Propulsion – Emphasis on nonnuclear components involving new mate-rial and control techniques for nuclear rocket engines.

(3) Electric Propulsion - Basic studies of electric field theory and propulsion devices involving electrostatic forces.

(4) Propellant Flame-Radiation studies to measure flame radiation temperatures and heat transmission.

MATERIALS RESEARCH:

(1) High Temperature Materials -- Research in high temperature material for rocket engines.

(2) Space Environment Effects on Materials -- Vacuum and radiation effects on polymeric materials.

NUCLEAR SCIENCES:

Radiation testing of rocket engine components.
Nuclear Mass Flow Device - to measure mass flow rates.

SPACE DYNAMICS:

Orbital transfer and rendezvous.
Interplanetary mission studies.
Perturbation studies.

AVIONICS

HIGH PERFORMANCE NAVIGATION SYSTEM (HIPERNAS II) - Complete guidance and navigation systems for strategic and tactical missiles, aircraft and aerospace vehicles, ship and submarine navigation and drone recovery.

ACCELEROMETERS AND DIGITAL VELOCITY METERS - The BAC III-B Linear Accelerometer has a range of $\pm45g$ and weight of 0.7 lbs. Combined with the external Digital Velocity Meter it yields a precision digital system whose pulse rate is proportional to the instantaneous acceleration.

RADIO RECEIVERS - Bell's 406- and 550 megacycles receivers meet the exacting requirements of missiles and guidance systems.

SPECIAL PRODUCTS AND SERVICES

HIGH-SPEED DATA PROCESSING - IBM 7090 computer and complete 1401 computer system.

MANUFACTURING RESEARCH -- Materials and processes modified and developed to meet specific and unusual requirements.

PRECISION MACHINING AND FABRICATION -- Manufacture and assembly of complex airframe and missile components.

TITANIUM FABRICATION -- Machining, hot forming and assembly of titanium parts.

MANUFACTURING SUBCONTRACTING -- Airframe and missile components including complete design, test and qualification.

HYPERSPEED PUMPS -- The design, manufacture and test of hi-pressure centrifugal pumps.

AIR CONVEYOR -- Provides frictionless platform for material handling.

PERSONALIZED LOAD CARRYING DEVICES -- Enables man to carry heavier loads with less fatigue over extended time periods.

LABORATORY CAPABILITIES

PROCESSES
Process development and specifications.
Vacuum Furnace.

CHEMISTRY
Inorganic, Organic, Physical and Analytical.
Solid and Liquid Propellants.

INSTRUMENTATION
Standards and Calibration.
Measurements.
Instrument development and evaluation.
Data acquisition and analysis
Human Factors

EQUIPMENT
Shock and vibration.   Electromechanics
Hydraulics.   Electronic noise
Static, acoustic and environmental test.

FLIGHT PERFORMANCE
Flight test and vehicle technology.

MATERIALS -- METALLIC AND NONMETALLIC
Ablative test and development.
Adhesive bonding evaluation and development.
Mechanical and thermal properties at -453 to 5000F.
Electron beam welding.
Coating evaluation.
High (<5000F) temperature oxidation tests.
Ceramic material development.

AUTOMATIC CHECKOUT EQUIPMENT -- A complete automatic checkout system developed for US Air Force missiles.

AIR TRAFFIC CONTROL BEACON EQUIPMENT -- Adds selective identification feature (SIF) to Mark X IFF equipment operating in conjunction with ground radar sets.

RADAR SYSTEMS -- Developed for both ground based and airborne applications including search, tracking, and seeker types.

BATTLEFIELD SURVEILLANCE SYSTEMS -- For target location, observation of troop movements and damage assessment utilizing reliable airborne sensors, positive position-reference equipment; data links, and precise ground sensor.

MISSILE AND DRONE RECOVERY SYSTEMS -- Successfully used for Regulus recovery. Combines features of the automatic landing system with Bell's secure command system.

SECURE TRANSMISSION SYSTEMS -- Designed for control, navigation, coded communication, and data transmission to offset countermeasures in electronic warfare.

AUTOMATIC FLIGHT CONTROLS -- A unique constant altitude hovering autopilot for Navy anti-submarine helicopters with special hydraulic servo valves, antenna drives and power systems.

AUTOMATIC LANDING SYSTEMS -- Available in either land or carrier based versions. The only ground-derived system available that affords precise and reliable aircraft control.

GYROSCOPES -- The Brig II gyroscope is a two-degree-of-freedom, floated instrument designed for aerospace applications where accuracy, small size, and light weight are essential.

Receivers, Transmitters, Coders, Beacons, Power Supplies, Electromagnetic and Electrostatic Research, RF Circuit and Microwave Equipment Development, Counter-measure and Counter-counter-measure Research, Analog and Digital Computation, and Data Processing Techniques.

ELECTRONICS RESEARCH -- Non-linear circuit theory; self adaptive filters; information theory and determination of optimum codes for pulse communication; polyphase frequency multipliers, multiple frequency pumping of parametric amplifiers, electromagnetic propagation in the atmosphere of the planets; consultation.

RADIO FREQUENCY INTERFERENCE -- RFI analysis of electronic systems, e.g., voice interference detection and analysis, measurement and analysis of communication systems. Detection, measurement and analysis of interference in RTT, pulse or radar systems. Automatic frequency measuring and monitoring equipment. Electromagnetic propagation theory development and field experimentation, antenna system development. Spectrum signature data collection and analysis. Theoretical RFI prediction techniques and mathematical modeling.

SERVICES -- Human factors analysis; studies and electronic simulation of man-machine interrelationships. Electronic range operation, data collection, data reduction and analysis.

1.4 Technology Development Time Line for the Bell-aerospace-company

1935
July -- Bell Aircraft Corporation founded by Lawrence D. Bell
Sept -- Bell receives first contract, to install engine in A-11 attack plane

1936
May -- Contract to build XFM-1 Airacuda

1937
Sept -- First flight of Airacuda

1938
April -- First flight of XP-39

1940
May -- First flight of XFL-1 Airabonita, for the Navy

1941
Jan -- Airacobra breaks world's power dive record at 630 mph
Sept -- Contract for three XP-59's, America's first jet aircraft
Nov -- Development of helicopter began
Dec -- Bell Aircraft chosen to build 668 B-29 bombers

1942
Oct -- First flight of XP-59 jet aircraft

1943
July -- First formal flight of helicopter
Dec -- First B-29 came off the assembly line

1944
April -- First flight of the all plywood XP-77 fighter

1946
Dec --First power flight of X-1

1947
Oct -- Bell X-1 breaks the sound bearer

1949
Jan -- Bell X-1 climber to 23,000 ft.
Mar -- Bell helicopter sets altitude record of 18,500 ft.
Mar -- Bell helicopter sets speed record of 133.9 mph
Aug -- Inauguration of first helicopter air mail service

1950
Y.S. Air Force announces Bell built Tarzan bomb

1951
April -- Bell's XV-3 convertiplane (tilt rotor VTOL) development contract
May -- Ground broken for new helicopter plant near Fort Worth, Texas

June -- Announcement of flight of jet powered X-5 -- the first airplane with adjustable wing angles

1952
Aug -- First Bell built nacelle for B-47 leaves assembly line

1954
Nov -- First flight of Bell developed VTOL airplane

1955
Feb -- Convertiplane first shown to public
March -- Bell announces manufacture of "Nike" engines
Sept -- XV-3 convertiplane flown for the first time

1957
Feb -- Initial flight test phase of X-14 VTOL completed
Nov -- GAM-63 Rascal missile turned over to Air Force. First operational squadron to use Bell developed missile

1959
May -- Project Mercury -- the first manned satellite program -- reaction controls will be developed by Bell
Oct -- Army's Sergeant ballistic missile will employ Bell's highly precise Accelerometer

1960
Sept -- Liquid fluorine and liquid hydrogen use as propellants in the first full scale firing of a complete turbo pump fed rocket engine
Dec -- Introduced full scale model of D188A VTOL fighter bomber

1961
Feb -- Put on public display was the Bell developed air cushion vehicle
June -- Built and tested rocket belt
Nov -- Development of a zero gravity belt was announced

1962
Jan -- Bell will build a 62 ft hydro skimmer
Feb -- Hip pack development announced
Sept -- V-22A VTOL airplane build contract received

1963
Jan -- Hip pack used on Mt. Everest climb
July -- Received contract to develop ascent rocket engine for NASA lunar excursion module
Oct -- Fluorine propulsion ready to apply to space missions

1964

April -- Fluorine rocket created 40,000 lb thrust, at Bell Test Center

May -- Lunar landing research vehicle had testing initiated

Nov -- Mariner III and IV missions to Mars, had their Agena rockets function perfectly

1965

Feb – Bell Agena-B rocket engine sent Ranger 8 on a successful picture taking mission to Mars

-- Pan American Airline 707's, flying the transatlantic route, selected Bell model VII-B accelerometers for their SGN-10 gyroscopes

May – The tri-service X-22 V/STOL aircraft was formally introduced

Aug – Minuteman III employs a post boost propulsion system for attitude and velocity control. Small liquid rockets will be designed and developed by Bell

Sept – To train Gemini 6 astronauts in rendezvous maneuvering, Bell developed and delivered a GEOS simulator for NASA

1966

March – Two new one man flying devices – the flying chair and the pogo stick – Were announced

April – Air cushion aircraft landing gear development was announced

Oct – To market the ACV, Bell Aerosystems Canada was created

Nov – SK-5 ACV's, built by Bell, began Navy service in Viet Nam

Dec – A record was set when the Bell Lunar Landing Research Vehicle stay aloft for 9 ½ minutes

1967

Mar – For the first time, the Bell X-22A VTOL aircraft took off vertically and transitioned to conventional flight

-- A 235 mile trip was made across ice clogged Lake Erie by an Sk5 ACV, for the first time

-- To train astronauts for landing on the moon, Bell will develop three lunar landing training vehicles

April – An Anchorage Alaska company made the first commercial purchase of two SK-5 ACV's

June – The one man rocket belt technology was modified for a two man rocket pogo that can carry a passenger or equipment

Aug – An air cushion landing gear was successfully flown

Nov – To simulate flight by a man on the moon or in orbit, Bell designed and built a

one man propulsion research apparatus (OMPRA)

Dec – Bell ascent engine for Apollo Lunar Module sent to Grumman

1968

June – High temperature alloys are being evaluated for application to advanced rocket and jet engines

Aug – Successful flight of pogo stick, which is maneuvered by leaning in direction of flight

Oct – Apollo 7, which utilizes 23 positive expulsion Bell tanks, was launched into orbit for 11 days

1969

June – New AERCAB program (Air Force Air Crew Escape and Rescue Capability) The goal is to construct and flight test an experimental pilot self-rescue system to evaluate the feasibility of utilizing a jet powered parawing to help pilots maneuver ejection seats away from hostile territory

Sept – Aircraft, with air cushion landing system, succeeded with first take off and landing from water

1970

Jan – AERCAB starts flight testing

March – The 100th flight was recorded by Bell's lunar landing training vehicle

June – Bell announced the world's only Mach 3.0 rotating arm erosion apparatus for evaluating the resistance of high strength aerospace materials to rain, sand, and ice particles – at sustained supersonic speeds

June – A contract was received to study the feasibility of employing Bell's air cushion landing gear to the space shuttle booster and orbiter vehicles

July – The 10th successful space flight, aboard an Agena space vehicle, was recorded for the miniature electrostatic accelerometer (MESA)

Sept – Twenty three flight tests, of AERCAB pilot self rescue system, were successful

Nov – Contract awarded for design, installation, and test of air cushion landing gear on De Havilland CC-115 airplane

-- U.S. Naval Oceanographic Office awards a contract for a shipboard gravity measurement system

1971

March --Bell selected for the design, construction, and test of two experimental 160 ton, 50 knot air cushion LC-JEFF(B) amphibious assault landing craft, by U.S. Navy

-- Bell's 100-ton surface effect test craft christened at New Orleans

1971

Aug – AN/SPN42 automatic landing system contract was awarded for installation on the USS Dwight D. Eisenhower

Oct – Bell joins a team for a NASA contract to design and construct two jet powered STOL transport research aircraft

1972

Feb – Received an Air Force contract to study the feasibility of employing an air cushion landing system for the recovery of unmanned aircraft

-- Work continued on using fluidized powder rocket propellants, when Bell received a contract from the Air Force Rocket Propulsion Laboratory

March – A SEV arctic operations study contract was awarded to the New Orleans Operation, by the Navy

June – Improving finite element methods of fracture analysis is the goal of a contract from the Air Force Flight Dynamics Lab

Oct – Development of a Marine remote area approach and landing system (MRAALS) contract was awarded to Bell by the U.S. Navy Electronic System Command

Nov – Contract for a remote land based automatic landing system, awarded by the Navy

1973

Feb – A Voyager air cushion vehicle begins operations in arctic weather, in the Northwest Territories

March – A world record is set as, for the first time, SES 100B hits speed greater than 70 knots

April – Advanced metallic structures are being studied under a USAF contract

May – The erosion test facility was successfully used to test space shuttle materials

-- All three Skylab spacecraft utilize Bell positive expulsion tanks

June – Bell completed NASA shipboard experiments with position locator aircraft (PLACE) system

July – The Wheatfield plant and the Bell Test Center were upgraded, in preparation for upcoming high energy laser projects

Oct – The 300th space launch, by Bell's Agena rocket engine, was recorded this Month

1974

April – The accidental discovery of the ice breaking capabilities of a Voyager ACV, was announced by the Canadian Ministry of Transport

1975

May – NASA's Viking program – to place an unmanned vehicle on Mars in 1976- received final delivery of Bell accelerometers

Jan – An extendable nozzle exit cone (ENEC) was successfully demonstrated

July – The Apollo spacecraft employed 33 Bell built positive expulsion tanks during the Apollo-Soyuz mission; the tanks performed perfectly

Aug – To assist in touchdown, Bell model X1 accelerometers were used on the Viking Mars Lander missions

1976

Feb – Bell is utilizing rocket technology in the search for ways to make better use of the nation's chief energy resource – coal. Bell investigated the feasibility of creating a small high mass flux gasifier, to convert coal to gas

1977

March – A borehole gravity meter (BHGM) was developed for the oil industry using Bell's accelerometers. To search for oil, it will be lowered into abandoned dry-hole well.

Nov – A high mass flux gasifier successfully completed its first tests.

1978

Jan – Ice breaking trials, with a LACV-30, were undertaken at Peoria, Illinois. Ice, as thick as ten inches, has been broken

June -- A Bell Agena rocket was used to launch a SEASAT-A ocean survey satellite

July – Bell delivered the JEFF(B) advanced development air cushion vehicle to the NAVY

Aug – The 14th AN/SPN-42(A) automatic carrier landing system was delivered for the new U.S. NAVY carrier, the USS Vinson

-- The ocean thermal energy conversion (OTEC) area contract was awarded to Bell. It is for a cold water pipe (CWP) model feasibility program

Oct – Bell-Halter SES – Americas first commercial surface effect ship – was launched. This is the first from Bell Aerospace Textron-Halter Marine Inc. joint venture

Nov – A contract is imminent for the acceleration measurement system for the NASA Galileo (Jupiter probe) spacecraft.

-- A solid propellant rocket was employed to successfully demonstrate the operation of a gas deployed nozzle skirt

Feb – Bell awarded a contract for development of a coal gasification system by N.Y. State Energy Research and Development Authority

-- Agena rocket engines have been successfully launched 300 times

-- An Air Force program was started: engineering development of a single delta oxygen generator and hydrogen pumped iodine laser

April – A TRADOC combined arms test activity (TCATA) program was fully Funded

Aug – The Peace Courage program received contracts for $44 million

Sept – Bell to develop high pressure fuel tanks, from a new improved titanium alloy for use on space vehicles. Supported by the Air Force Rocket Propulsion Lab

-- The U.S. Army Missile Command provided funding for Phase I of a chemical laser nozzle technology program

-- Initial funding received for the advanced developed model of the gravity gradiometer

--Rough water trials were conducted with a Bell-Halter SES. In the Gulf of Mexico, an average of 12 foot waves were encountered, but some waves were over 20 feet high

-- Received a $10 million software system contract, from Grumman, for a dynamic electromagnetic environment simulator (DEES) that will be used by the Air Force

-- NASA honored Nelson Roth, chief engineer of combustion devices, for his invention "Hot Gas Regeneratively Cooled Propulsion Engine"

Oct – Contract received for a Space Lab-3 MESA (Miniature Electrostatic Accelerometer)

April – Design and fabrication of three flight weight laser hardware modules were demonstrated by the Lambda High Energy Laser Program

May – Dalmo Victor Operations, at Grants Pass Oregan, began testing of the miniaturized radar warning (Mark III) system at the naval test facilities

July -- Design and development of a new automatic carrier landing system (SPN/XX) will be pursued under a $12.5 million multiyear U.S. Navy contract. The system will replace the 21 U.S. Navy SPN/42 carrier landing systems Bell had designed, built, and installed earlier

July – Dalmo Victor Operations will study radio frequency identification sensors, (RFIS) under a nine month contract from Air Force Wal's Reconnaissance and Weapons Delivery Division

Sept – Bell will analyze electronic warfare communication equipment under a two year contract from the U.S.A.F. Electronic Warfare Center.

-- Dalmo Victor Operations will study a New Threat Warning System (NTWS) for the 1990's, under a U.S. Air Force Contract

-- Bell will study integrating on-board sensors with crew needs on armored vehicles for a Vehicle Integrated Defense System, under a U.S. Army Tank Automotive Command contract

Dec – Bell will continue to provide engineering and operation services for the Electromagnetic Environment Test Facility, under a new five year contract

-- Martin Marietta awarded a three year contract for full scale development of a Shock Isolation Unit (SIU) to be used on MX missile launcher

Jan – Contract received, from the Army, for engineering, software development, and fabrication of an Automatic Target Control System (ATCS)

Mar – Bell's New Orleans Operations received a Navy contract for support of the Amphibious Assault Landing Craft

April – Under an Air Force contract, Bell will study new Alternate Aircraft Takeoff Systems (AATS) to obtain greater mobility of fighter planes from bomb damaged runways

-- The first ABCAS (Active Beacon Collision Avoidance System) was delivered to the FAA, by Dalmo Victor Operations

-- As part of the World-Wide DoD Radar Tracking Network, the System Test Facility successfully supported the maiden flight of the NASA space shuttle Columbia

June – A $40 million contract, won by the New Orleans Operation, was for a new type of amphibious landing craft, to be called the "LCAC" (Landing Craft, Air Cushion). Included are two options for future production of six LCAC's, one option for $81 million and the other for $51 million

-- Bell's Ft Story facility for assembly and test of the LCAV-30 became Operational

-- Collision warning systems – TCAS I for general aviation and TCAS II for general aviation – was recommended by the Federal Aviation Administration. TCAS II includes a scanning directional antenna and cockpit display

July – The Chemical Oxygen Iodine Laser (COIL) program received an additional $1.6 million from the U.S. Air Force, for additional testing

Sept -- The preliminary design of an air cushion landing gear, for a new type of triphibious aircraft that can land or take off from land, water, or snow, will be undertaken under a new contract foe Bell Aerospace Canada

Nov – A contract was received for a Low Altitude Defense (LoAD) interceptor control system for defense of U.S. land-based intercontinental ballistic missiles

1982

Jan – A Gravity Sensors System (GSS), the only one in the world, was developed for the U.S. Navy and installed aboard the USNS Vanguard. This will improve the survey of inertial navigation systems and fine grain mapping of the earth's gravity field

Feb – More than 354 successful space flights have been completed by Bell's Agena rocket, since it was introduced 23 years ago

Mar – Niagara Frontier Operations received a contract, from Raytheon, for a U.S. Navy EHF satellite communications program for a stabilized antenna system (NESP)

-- Niagara Frontier Operations was awarded a U.S. Army contract to construct a laser test facility at the White Sands Missile Range in New Mexico

April – The New Orleans Operation was requested, by the U.S. Navy Atlantic Fleet, to evaluate ACV's for mine countermeasure activities – a new mission for ACV's

May – The "LoAD" program had its name officially changed to "Sentry", by the U.S. Army

June – Two Piedmont Airline's Boeing 727's were used to evaluate two of Bell's Beacon-based Collision Avoidance Systems (BCAS), for the first phase of FAA tests in an operational environment; 900 hours of flight testing were successfully completed

-- Raytheon awards Niagara Frontier Operations a six month for a MILSTAR trade study

-- Dalmo Victor Operations successfully flight tested a TCAS II feasibility model with a 90^0 sector transmission antenna, on a Boeing 727 in New York City, Boston, and Chicago

-- A 16-month contract was won by the Niagara Frontier Operations for conceptual design, and sustained engineering for the LVT(X) (Landing Vehicle, Tracked Experimental)

July – A U.S. Air Force feasibility study contract, for a proposed deep based weapons system, was awarded to the Niagara Frontier Operations and Gilbert/Commonwealth team. Niagara Frontier Operations will act as program manager and will also study the missile transporter/erector/director systems

Sept – Bell won a contract to enhance the Advanced Electronic Warfare Evaluation Display System/Interactive Research System (AEWEDS/IRS) at the A. F. Electronic Warfare Center, at Kelly Air Force Base

--The U.S. Army Electronic Laboratories awarded a contract, to Dalmo Victor Operations, for 23 Engineering Development Model AN/APR-39A Radar Warning Systems (Mark III) and 11 flight line test sets on a joint program for the U.S. Marine Corps, the U.S. Army, and the Canadian Air Force

Oct – Bell received a contract for an AFGL/USAF Gravity Gradiometer Survey System (GGSS), utilizing Bell's precision accelerometers and advanced inertial system technology; it will be tested on an aircraft or a land vehicle

-- The first two production Surface Effect Ships, the Sea Hawk and the Shearwater, were delivered to the U.S. Coast Guard by Bell-Halter

Dec – Bell made final delivery, of the Peace Courage hardware, to a NATO country

1983

Jan – Dalmo Victor Operations and Sperry formed a joint venture to manufacture and market the TCAS II – an airborne collision avoidance system for commercial aircraft

Mar – A high resolution Accelerometer package that employed Bell's Model XI accelerometer, flew on the maiden voyage of the Space Shuttle Challenger. Bell Technical Operations helped track the Challenger

April – Two new types of U.S. Navy craft will be designed by the New Orleans Operation: The Patrol Boat Multi Mission (PBM) and the Minesweeper Hunter (MSH). Bothe crafts will be designed as surface effect ships

Nov – Work began, at the Niagara Frontier Operations, on a N.A.S.A. and Marshall Space Flight Center contract for high performance alloy electroforming. A Bell developed electroformed alloy will be evaluated for its material property advantages: high temperature strength and ductility. These properties would allow a reduction in engine weight and cost. The material would be used in the structural jacket of the Space Shuttle's main engine

1984

Jan -- JEFF(B), Bell's air cushion vehicle, successfully completed 20.2 hours of simulated Arctic testing

Feb – Five Three Axis Instrumentation (TAI) accelerometers were ordered by the Charles Stark Draper Laboratory, for use on Trident II test flights

Mar -- A subcontract was won by the Niagara Frontier Operations for the development of an area reprogramming capability (ARC)
-- Four major aerospace companies, including Bell, won design competition contracts for a hard mobile launcher (HML) for the small intercontinental ballistic missile (SICBM)

July -- Niagara Frontier Operations conducted a flight test program with the AN/SPN-46(V) automatic carrier landing system using a Navy A7 aircraft. Bell has been the sole supplier of the landing systems to the U.S. Navy since 1960. Fourteen carriers now have the system and a system has been delivered to the fifteenth

Nov -- Textron Inc. announced the reorganization of Bell Aerospace Textron, effective January 1, 1985. Dalmo Victor Textron, a new Textron Inc. unit will be established; it will be composed of Dalmo Victor Operations, Belmont, Ca; EM Systems Inc., Fremont, Ca., and Bell Technical Operations, Tucson, Arizona. Bell Aerospace Textron will continue to include Niagara Frontier Operations, Buffalo, New Orleans Operations; Bell Halter Inc., New Orleans; and Bell Aerospace Textron, Grand Bend, Ontario
-- A new class of coastal minesweepers, the Minesweeper Hunter (MSH-1), will be developed for the U.S. Navy by the New Orleans Operation using air cushion vehicle technology

Dec -- Dalmo Victor Operations received a contract to develop and produce active electronic buoys (AEB), a new expendable system designed to protect naval vessels
-- An original video compression (VIDCOMP) system, that is capable of transmitting compressed television images from a military scout vehicle or aircraft via a tactical frequency-hopping radio, was created by Dalmo Victor Operations. The system can relay images on a secure channel nearly Instantaneously

1985

Jan -- Rubber skirts, for Bell's air supported craft, will be manufactured by a joint venture with Avon Industrial Polymers Ltd. of Britain
-- Bell's HAS (Hydrazine Actuation System) and Peace Courage are operational. This is the only mono-propellant post boost propulsion system in the Free World

-- The 100 ton LCAC-1 has exceeded all contract specifications for speed, range, and payload; 11 more will be built

Feb -- The engineering development model of AN/SPN-46(V) was successfully flight tested

April -- A HAS (Hydrogen Actuation System) Mark II gas generator will be produced, under contract

May -- A contract was awarded to design, build, and test a candidate auxiliary propulsion rocket thruster for NASA's space station.

June -- A subcontract was received to develop a cooled mirror for the Los Alamos National Laboratory. It will be applied to a free electron laser. This is Bell's first contract that is related to the Strategic Defense Initiative

July -- For the U.S. Army Benet Laboratory, Bell will test combustors of liquid propellant guns

Aug -- In support of Boeing, Bell will design, study, and test a Sagittar (Space Experiment Control System) liquid propulsion system
-- Bell will work on an extendable nozzle for a solid propellant rocket booster, for Thiokol, as part of the Kinetic Energy Weapons System for SDI (Strategic Defense Initiative)

Sept -- Development of an air cushion crash rescue vehicle, will Niagara Frontier Operations
-- Niagara Frontier Operations received contracts, for liquid gun feed and propellant handling systems, from the Army Watervliet Arsenal and Ballistic Research Laboratory

Nov -- Ford Aerospace Aeronutronics Division awarded a contract to Bell for concept definition and design studies for Sagittar (Space Experiment Control System) liquid propulsion system

Dec -- Lt. General James Abrahamson selected the space based rocket launching KEW (Kinetic Energy Weapon) -- which Bell helped develop -- as the preferred concept, over the electromagnetic launched concept
-- Bell Niagara Frontier Operations and Magnavox were awarded a military contract to provide engineering development and sophisticated electronic systems, which includes 15 antennas and pointing systems, for the SCOTT system (Single Channel Objective Tactical Terminal). SCOTT is part of the Tri-Service MILSTAR terminal system which also includes the U.S. Navy NESP (Navy Extra-high-frequency Satellite-communication Program), in addition to the U.S. Air Force MILSTAR program

1.5 Samples of Bell Company-Sponsored Individual Research and Development (IR&D) Projects

To effectively utilize its limited resources, Bell created specific objectives that had to be met by its IR&D projects:

"(1) To determine the applicability of scientific discoveries to the products of current interest, and to determine the requirement for further research efforts

(2) To perform the research, and exploratory or advanced development, that will provide the technology which is necessary for the orderly Design and development of products of current or prospective interest

(3) To advance the state-of-the-art within selected fields of specialization

(4) To conduct detail studies, on a continuing basis, of opportunities presented to the company for application of its professional competenc and supporting facilities in the development of new products of prospective interest to the government."

BELL AEROSYSTEMS COMPANY IR&D PROJECTS FOR 1967

PART I – RESEARCH

SECTION I – GENERAL RESEARCH PROJECTS
PROJECT 1: Advanced Composite Materials and Structures Research
PROJECT 2: Learning Nets and Advanced Adaptive Control Techniques
PROJECT 3: Advanced Microminiaturization Techniques
PROJECT 4: Target Acquisition and Weapons Delivery Methods

SECTION II – PROPULSION RESEARCH PROJECTS
PROJECT 5: Propulsion Technology Research
PROJECT 6: Advanced Turbopump Fed Engine
PROJECT 7: Advanced Auxiliary Rocket Engine
PROJECT 8: Electric Thruster for Space Application
PROJECT 9: Advanced Positive Expulsion Techniques and Materials

SECTION III – AIRBORNE VEHICLE RESEARCH PROJECTS
PROJECT 10: –Air Cushion Vehicle Systems
PROJECT 11: Small Lift Devices

SECTION IV – SPACE RESEARCH PROJECTS
PROJECT 12: Investigation of Advanced Space Systems

SECTION V – COMMUNICATIONS RESEARCH
PROJECT 13: Advanced Communications Research

SECTION VI – V/STOL RESEARCH PROJECTS
PROJECT 14: V/STOL Aircraft Research

PART II – DEVELOPMENT

SECTION I – PROPULSION DEVELOPMENT PROJECTS
PROJECT 1: Advanced Prepack Technology

SECTION II – LANDING SYSTEMS DEVELOPMENT PROJECTS
PROJECT 2: Development of an Advanced Landing System

SECTION III – INERTIAL SYSTEMS DEVELOPMENT PROJECTS
PROJECT 3: Inertia Instrument Development
PROJECT 4: Development of a Low Weight High-Performance Inertia Navigation System – Hipernas III
PROJECT 5: Fire Control System Development

SECTION IV – TARGET LOCATOR DEVELOPMENT PROJECTS
PROJECT 6: Advanced Command and Control Systems
PROJECT 7: Visual Airborne Target Locator System

BELL AEROSPACE COMPANY IR&D PROJECTS FOR 1970

PART I – RESEARCH

1: Composite Materials and Structures
2: Advanced Materials and Mechanics of Materials Research
3: Propulsion Materials Research
4: Combustion Instability Analysis
5: Propulsion Fluid Mechanics and Heat Transfer
6: Liquid Auxiliary Rocket Engine
7: Gaseous Auxiliary Rocket Engine
8: Upper Stage Engine
9: Air-Augmented Propulsion System Technology
10: Advanced Positive Expulsion Techniques

15

1.6 Bell-aerospace-company Facilities in Western New York

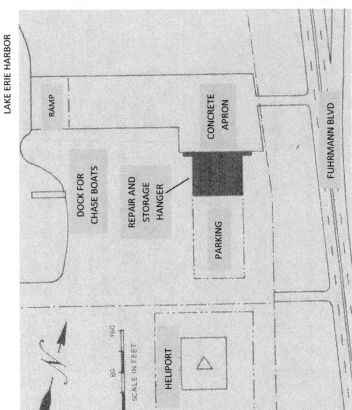

LARRY BELL OFFICE

1.6-b Entrance to Bell Aerospace Division of Textron, Wheatfield Plant

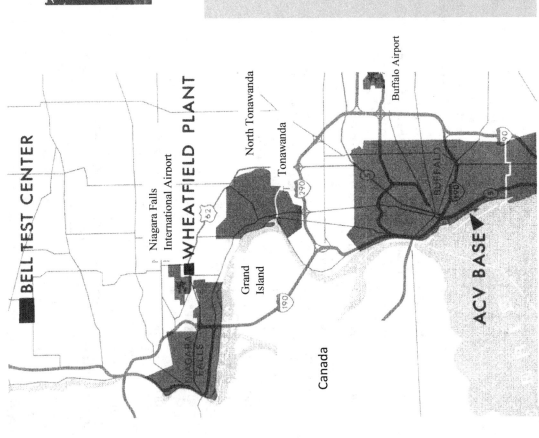

LAKE ERIE HARBOR

RAMP

DOCK FOR CHASE BOATS

REPAIR AND STORAGE HANGER

CONCRETE APRON

PARKING

HELIPORT

SCALE IN FEET
0   80   160

FUHRMANN BLVD

1.6-c ACV Base for Testing Full Size Air Cushion Vehicles

BELL TEST CENTER

WHEATFIELD PLANT

Niagara Falls International Airport

North Tonawanda

Tonawanda

Grand Island

NIAGARA FALLS

Canada

Buffalo Airport

BUFFALO

ACV BASE

1.6-a Map Showing Location of Bell Facilities in Western New York

16

1.6-d Aerial Photograph of the Bell Wheatfield Plant

the Wheatfield Plant

BELL AEROSYSTEMS COMPANY POST OFFICE BOX NO. 1
BUFFALO, NEW YORK 14240

NIAGARA FALLS BLVD. (U.S. HIGHWAY 62) TOWNSHIP OF WHEATFIELD
NIAGARA COUNTY, NEW YORK

The headquarters of the Bell Aerosystems Company is located in the building complex known as the Wheatfield Plant, named for the township in which it is located. Here are functions relative to Administration, Finance, Manufacturing, Engineering, Marketing and Planning. Wheatfield also provides extensive offices and la boratories for the research, development and production programs of Bell's Engineers and Scientists. Here too is located th e principal manufacturing effort together with the vital product assurance and testing functions. A separate building houses the Electronic Data Processing Center equipped with the latest in digital and analog computers, and simulation equipment.

The Wheatfield Plant occupies in excess of eighty acres on the north side of Niagara Falls Blvd. Addi tional acreage across the Blvd., nearly duplicates this area and provides parking for approximately 4,000 cars and room for future facility expansion. The basic property lease runs through mid 1973 with portions under lease through early 1979. Air freight moves easily in and out of the adjacent Niagara Falls International Airport. Truck traffic utilizes U.S. Highway 62 or N.Y. 18 and six spurs of the Penn-Central Railroad serve the plant.

There are hangars and additional buildings for ACV and aircraft modification and test and for cus tomer training and product servicing. These and still others, that may be noted on the plot plan, are all well equipped and protected against fire and other hazards.

1.6-e   Layout of the Bell Wheatfield Plant

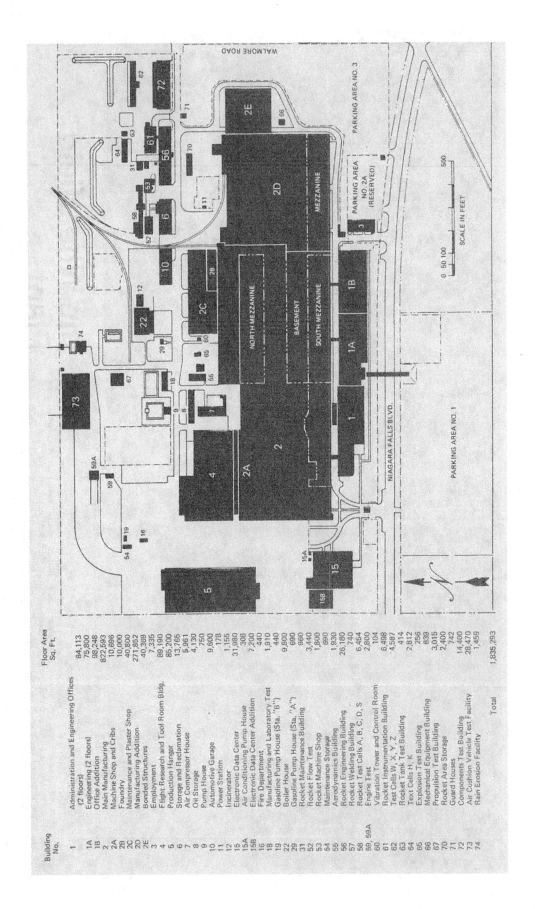

Building No.		Floor Area Sq. Ft.
1	Administration and Engineering Offices (2 floors)	84,113
1A	Engineering (2 floors)	75,800
1B	Office Addition	98,248
2	Main Manufacturing	822,593
2A	Machine Shop and Cribs	10,696
2B	Foundry	10,000
2C	Maintenance and Plaster Shop	40,800
2D	Manufacturing Addition	271,852
2E	Bonded Structures	40,369
3	Employment	7,335
4	Flight Research and Tool Room Bldg.	89,190
5	Production Hangar	86,200
6	Storage and Reclamation	13,765
7	Air Compressor House	5,961
8	Oil Storage	4,130
9	Pump House	750
10	Automotive Garage	9,600
11	Power Station	178
12	Incinerator	1,155
15	Electronic Data Center	31,980
15A	Air Conditioning Pump House	308
15B	Electronic Data Center Addition	7,200
16	Fire Department	440
18	Manufacturing and Laboratory Test	1,910
19	Gasoline Pump House (Sta. "B")	440
22	Boiler House	9,800
29	Gasoline Pump House (Sta. "A")	690
31	Rocket Maintenance Building	960
52	Rocket Flow Test	3,440
53	Rocket Machine Shop	1,800
54	Maintenance Storage	690
55	Aerodynamics Building	1,930
56	Rocket Engineering Building	26,180
57	Rocket Welding Building	740
58	Rocket Test Cells A, B, C, D, S	6,454
59, 59A	Engine Test	2,800
60	Vibration Tower and Control Room	104
61	Rocket Instrumentation Building	6,498
62	Test Cells W, X, Y, Z	4,587
63	Rocket Tank Test Building	414
64	Text Cells H, K	2,812
65	Explosion Test Building	266
66	Mechanical Equipment Building	639
67	Propulsion Test Building	3,015
70	Rocket Area Storage	2,400
71	Guard Houses	742
72	Components Test Building	14,400
73	Air Cushion Vehicle Test Facility	28,470
74	Rain Erosion Facility	1,459
	Total	1,835,293

18

1.6-f   Bell Test Center, Town of Porter -- Balmer and Porter Center Roads

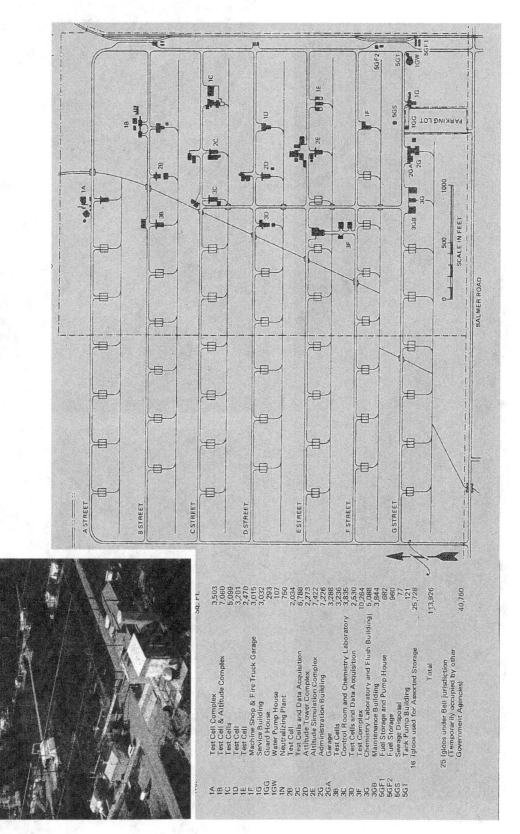

		SQ. FT.
1A	Test Cell Complex	3,503
1B	Test Cell & Altitude Complex	7,060
1C	Test Cells	5,099
1D	Test Cell	3,201
1E	Test Cell	2,470
1F	Machine Shop & Fire Truck Garage	3,015
1G	Service Building	3,032
1GG	Guard House	293
1GW	Water Pump House	107
1N	Neutralizing Plant	750
2B	Test Cell	2,034
2C	Test Cells and Data Acquisition	6,788
2D	Altitude Tower Complex	2,273
2E	Altitude Simulation Complex	7,422
2G	Administration Building	7,226
2GA	Garage	3,288
3B	Test Cells	3,236
3C	Control Room and Chemistry Laboratory	3,835
3D	Test Cells and Data Acquisition	2,530
3F	Test Complex	10,264
3G	Chemistry Laboratory and Flush Building	5,088
3GB	Maintenance Building	3,844
5GF1	Fuel Storage and Pump House	682
5GF2	Fuel Storage	960
5GS	Sewage Disposal	77
5GT	Tank Pump Building	121
16 Igloos used for Assorted Storage		25,728
	Total	113,926
25 Igloos under Bell jurisdiction		40,750
(Temporarily occupied by other		
Government Agencies)		

A STREET
B STREET
C STREET
D STREET
E STREET
F STREET
G STREET

BALMER ROAD

SCALE IN FEET
0   500   1000

1A
1B
1C
1D
1E
1F
1G
1GG
1GW
1N
2B
2C
2D
2E
2G
2GA
3B
3C
3D
3F
3GB
3G
5GF1
5GF2
5GS
5GT

19

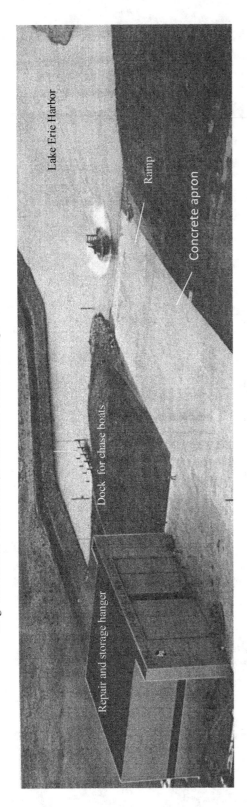

1.6-g Air Cushion Vehicle Base, Buffalo New York, for testing of full size air cushion vehicles

Typical test facilities at Bell Test Center, Town of Porter, New York

1.6-i Rocket engine vertical test stand

1.6-h High altitude simulation test stand

20

2.0   Air Craft

2.1 Conventional Airplanes

2.1.1 History of Bell Aircraft Corporation Airplane Development and/or Production

Airplane	Engine	Ceiling (ft)	Max Level Speed (mph)	Developed By	Units Built By Bell	No. OF Models	First Flight	Comments
YFM-1 Airacuda	twin engine pusher	30,500	277	Bell	13	4	9-28-39	Bomber interceptor and bomber. First airplane designed and built by Bell. Used innovative tricycle landing gear. One 37 mm cannon, remotely controlled, mounted in each engine nacelle.
P-39 Aircobra	propeller	20,000 (N & Q)	375 (N&Q)	Bell	9,584	16 (X,Y,A-Q)	4-6-38	Fighter. Used for ground attacks, by Russia. First production airplane with: tricycle landing gear, car-like cockpit doors, engine behind cockpit, cannon barrel through center of propeller hub.
XFL-1 Airabonita	propeller	30,900	307	Bell	1	1	5-13-40	Experimental shipboard interceptor developed for Navy. Identical to P-39, but used tail wheel instead of tricycle gear.
P-400	propeller	20,000	341 (loaded)	Bell	(675 modified)	1	4-1-41	Fighter. Modified P-39D. Cannon replaced by Hispano 20-mm cannon. Used in the South Pacific in ground-support role.
P-63 Kingcobra	propeller	24,100 prototype	421 prototype	Bell with Russian input	3,506	6	12-7-42	Fighter. Used to address P-39 problems. Installed better engine, a second supercharger, larger cowling panels for better access to armament, increased pilot armor, fuselage and under wing hard points, extra fuel tanks, moved cannon forward to increase cannon ammunition supply, four bladed propeller, larger wings, taller tail.
L-39	propeller			Bell	2	1	4-23-46	Experimental swept wing, fixed at 35 degrees. Modified P-63C. Used for wind tunnel testing.
XP-77	propeller	30,100	330	Bell	2	1	4-1-44	Mainly wood construction. Prototype only.
P-59 Aircomet	jet	46,200	413	Bell. Inspired by U.K jet program	68	5	10-1-42	First U.S. jet fighter. Designed and built during WWII. Never saw combat.
XP-83	jet	45,000	522	Bell	2	1	2-25-45	Escort jet fighter with extended range due to increased fuel.
F-7F Tigercat (modified)	twin engine propeller	40,400	460	Airplane: Grumman Control : Bell	Airplane: 12 Control :12	Airplane:1 R-control :1	3-15-46 (report)	Bell developed and installed a radio control system on a Grumman F-7F fighter. It was remotely controlled from the ground or from a Mother aircraft.
Bell X-1	rocket	90,000+ (X-1E) 74,700 (X-1A)	1,450 (X-1E) 1,620 (X-1A)	Bell	7	5	12-15-55 12-12-53	Experimental rocket plane. First aircraft to exceed the speed of sound in level flight. Maximum speed achieved Mach 2.21.
Bell X-2	rocket	126,200	2,094 (Mach 3.2)	Bell	2	1	6-27-52	Research aircraft. Study flight characteristics between Mach 2 and Mach 3.
Bell X-5	turbojet	49,900	716	Bell. Inspired by German P-1101 swept wing model (never flown)	2	1	6-20-51	Swept wings. First time wing angle is changeable in flight. Successfully demonstrated improved performance using swept wings.
B-29 Super fortress	four engine propeller	40,000	350	Boeing	668	1	9-21-42	Strategic long distance bomber.
MQM-57A Drone	twin propeller	40,400	460	Plane: Northrop Radio Control: Bell	2	1	1945	Small observation drone. Bell's ground radio control increased the drone's range and allowed multiple drone operations in same area.
Bomber	twin prop	-	-	Bell Concept	0	0	-	Proposal. No contract awarded. Probably model no. 10.
X-16 Bald Eagle	twin jet			Bell	0	0	-	Reconnaissance airplane. Mockup only. No production contract.
S-1	rocket	200,000+	Mach 3.2	Bell	1	1	-	Inhabited missile research. Supersonic/transonic research, very high altitude research.

Cenkner Table

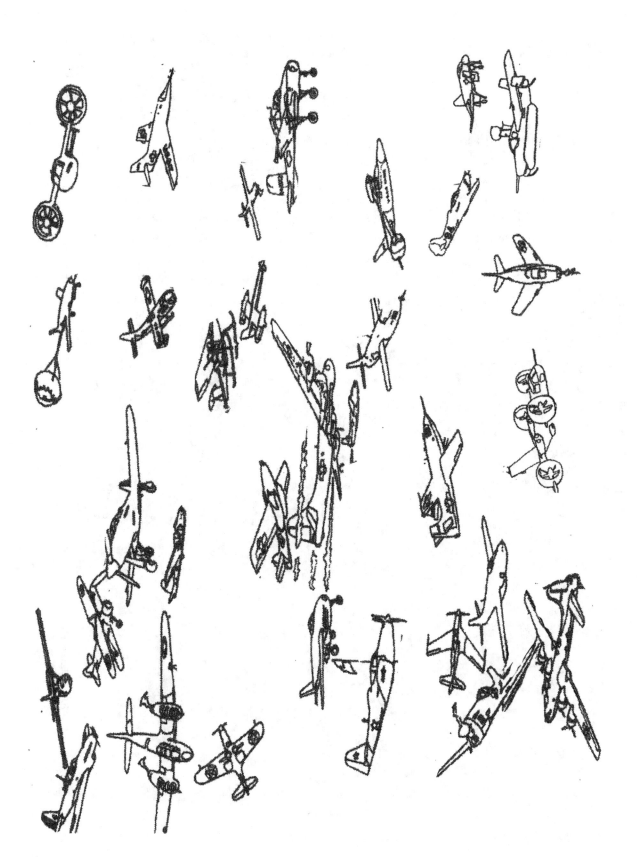

23

2.1.2 Airplane Development History Graph

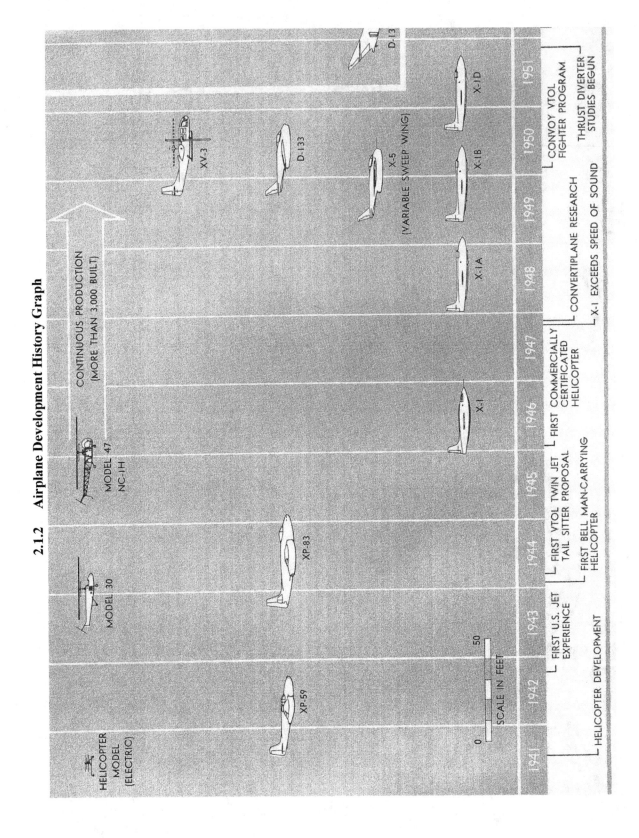

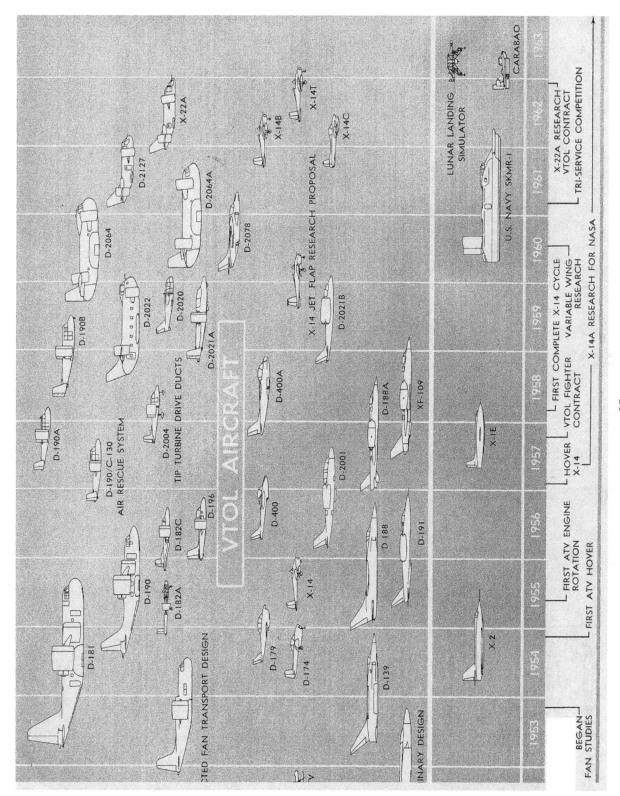

VTOL AIRCRAFT

25

2.1.3 Chronology of Major Bell Airplane "Firsts": 1938-1966

Year	Aircraft	Description
1938	XP-39	First Fighter Airplane in the United States to Utilize 37 mm Cannon Firing Through Hollow Prop Hub and Retractable Tricycle Landing Gear.
1942	P-59	First Jet-Powered Airplane in the United States.
1947	X-1	First Airplane in the World to Fly Faster than the Speed of Sound (670 MPH).
1949	Model 47	First Commercially Licensed Helicopter in the World.
1951	X-5	First Airplane in this Country Designed to Vary the Sweep of its Wings While in Flight.
1953	X-1A	Set Speed and Altitude Records of 1,600 MPH and 90,000 Feet.
1954	Air Test Vehicle	First Successful Jet V/STOL.
1955	Automatic Landing System	First Practical Automatic All-Weather Landing System in the United States.
1956	X-2	Set Altitude and Speed Records of 126,000 Feet and 2,148 MPH.
1956	Fluorine Rocket Propulsion System	First Practical Application to Rocket Propulsion of the High Energy Liquid Propellant Oxidizer, Fluorine.
1961	Rocket Belt	First Portable Rocket Propulsion System for Controlled Manned Flight.
1961	Mercury Reaction Controls	First Man-Rated, Space-Proven Attitude Control System.
1963	Hydroskimmer SKMR-1	Largest Air Cushion Vehicle Operative in the United States.
1965	Lunar Landing Research Vehicle	VTOL Test Vehicle to Simulate Lunar Landing in Earth Environment.
1965	SK-5	Nation's First Commercial Passenger Service Utilizing Air Cushion Vehicles.
1966	X-22A	Dual Tandem Ducted Prop Research VTOL Aircraft
1966	Air Cushion Landing Gear	First Application of Air Cushion to Replace Conventional Landing Gear

26

2.1.4 Bell's XFM-1 Airacuda Twin Engine Bomber Interceptor

2.1.4-a   XFM-1 Airacuda long range bomber destroyer. First airplane developed by Bell. Twin engine pusher. One 37 mm cannon and one gunner in each engine nacelle and 2-50 cal and 2-30 cal machine guns. Revolutionary tricycle landing gear.

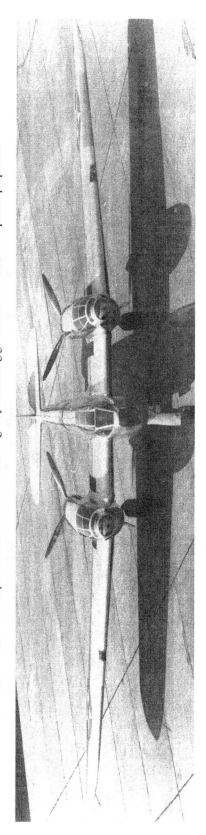

2.1.4-b   Front view of parked XFM-1 Airacuda showing tricycle landing gear and three bladed twin pusher propellers.

27

2.1.4-c Parked YFM-1 Airacudes

The YFM-1 (prototype, fighter, multiplace) was the first airplane designed and built by the Bell Aircraft Corporation. It was intended to be a long range bomber interceptor, capable of attacking bombers that were beyond the range of single seat fighters.

There were five crew members, with one in each of the engine modules. Their main job was to load 110 rounds of ammunition in each of the 37 mm cannons, but they could also aim and fire the cannons. Normally, this was the job of the fire control officer in the nose of the cockpit.

Most of the planes were built as tail draggers (above) but three of them were fitted with a tricycle landing gear.

28

2.1.5   Bell's P-39 Airacobra Fighter and Ground Attack Airplane

2.1.5-a   P-39 with revolutionary Bell tricycle landing gear – greatly improved pilot visibility while on the ground

2.1.5-b   P-39 taking off -- tricycle landing gear fully retracted to minimize aerodynamic drag

29

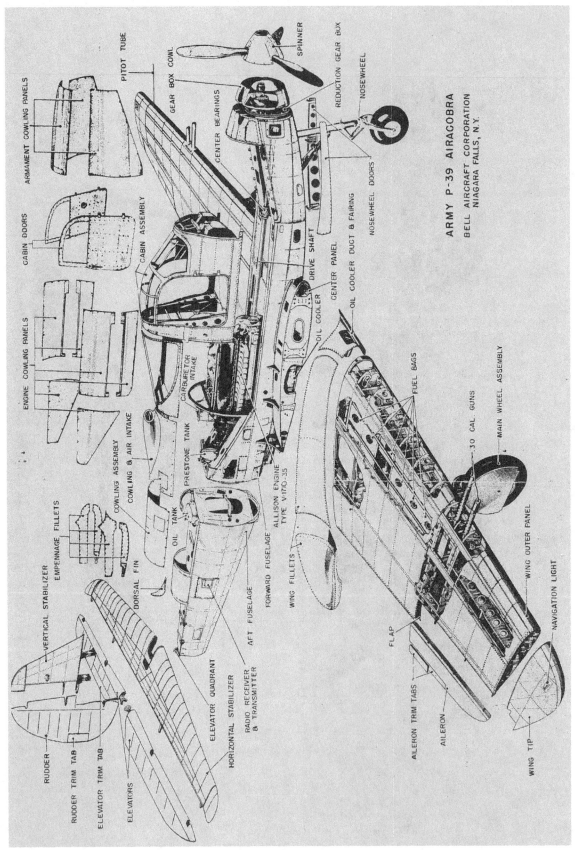

RUDDER
RUDDER TRIM TAB
ELEVATOR TRIM TAB
ELEVATORS

VERTICAL STABILIZER
EMPENNAGE FILLETS
DORSAL FIN

ELEVATOR QUADRANT
HORIZONTAL STABILIZER
RADIO RECEIVER & TRANSMITTER

ARMAMENT COWLING PANELS
PITOT TUBE

CABIN DOORS
CABIN ASSEMBLY

ENGINE COWLING PANELS

OIL TANK
COWLING ASSEMBLY
COWLING & AIR INTAKE
CARBURETOR INTAKE
PRESTONE TANK
ALLISON ENGINE TYPE V-1710-35
FORWARD FUSELAGE
AFT FUSELAGE
WING FILLETS

GEAR BOX COWL
CENTER BEARINGS

SPINNER
REDUCTION GEAR BOX
NOSEWHEEL
NOSEWHEEL DOORS

DRIVE SHAFT
OIL COOLER
CENTER PANEL
OIL COOLER DUCT & FAIRING

FUEL BAGS
.30 CAL. GUNS
MAIN WHEEL ASSEMBLY
WING OUTER PANEL
NAVIGATION LIGHT

AILERON TRIM TABS
AILERON
FLAP
WING TIP

ARMY P-39 AIRACOBRA
BELL AIRCRAFT CORPORATION
NIAGARA FALLS, N.Y.

30

2.1.5-c Exploded View of P-39

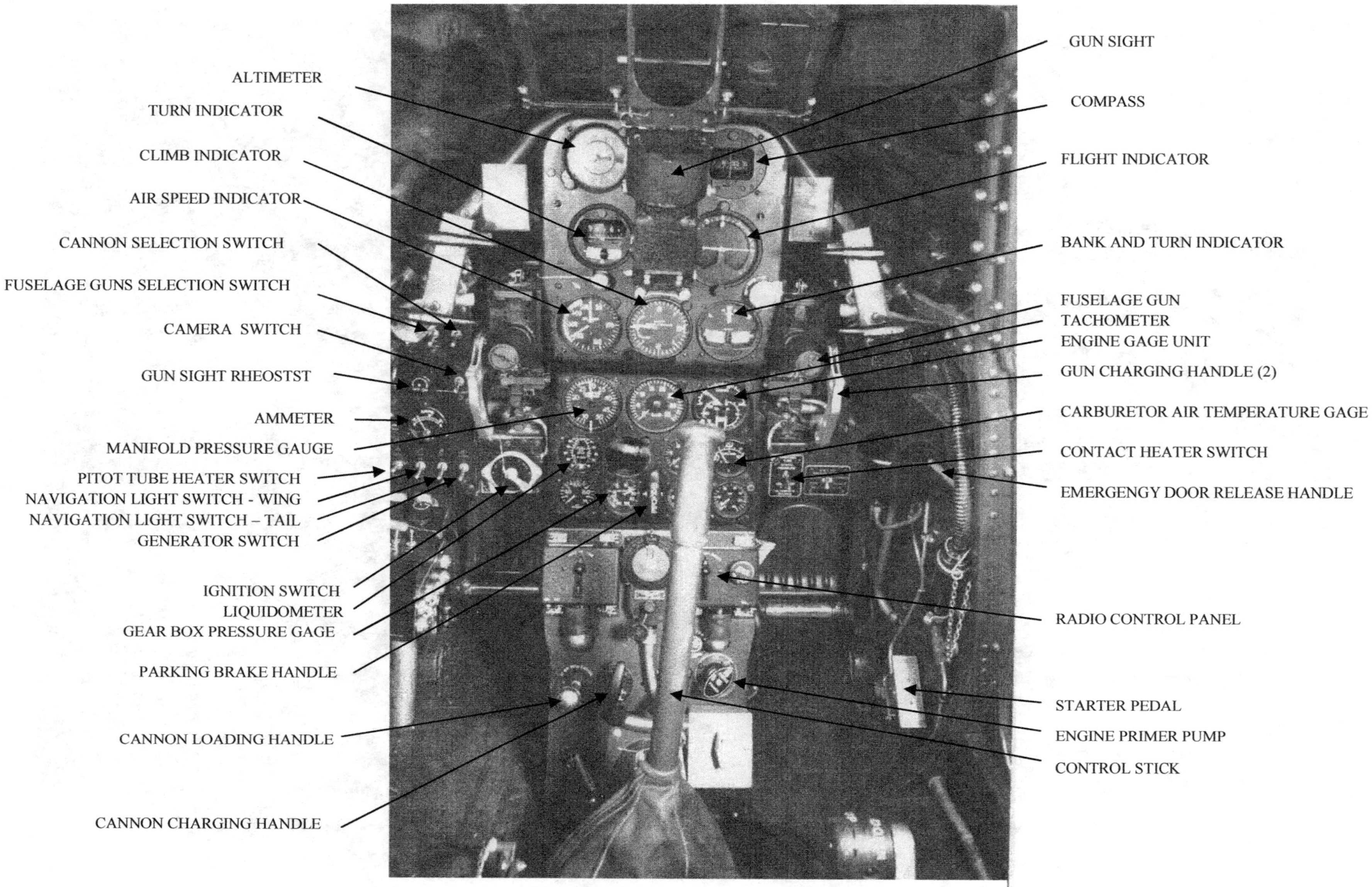

GUN SIGHT

COMPASS

FLIGHT INDICATOR

BANK AND TURN INDICATOR

FUSELAGE GUN
TACHOMETER
ENGINE GAGE UNIT

GUN CHARGING HANDLE (2)

CARBURETOR AIR TEMPERATURE GAGE

CONTACT HEATER SWITCH

EMERGENGY DOOR RELEASE HANDLE

RADIO CONTROL PANEL

STARTER PEDAL

ENGINE PRIMER PUMP

CONTROL STICK

ALTIMETER

TURN INDICATOR

CLIMB INDICATOR

AIR SPEED INDICATOR

CANNON SELECTION SWITCH

FUSELAGE GUNS SELECTION SWITCH

CAMERA SWITCH

GUN SIGHT RHEOSTST

AMMETER

MANIFOLD PRESSURE GAUGE

PITOT TUBE HEATER SWITCH

NAVIGATION LIGHT SWITCH – WING

NAVIGATION LIGHT SWITCH – TAIL

GENERATOR SWITCH

IGNITION SWITCH

LIQUIDOMETER

GEAR BOX PRESSURE GAGE

PARKING BRAKE HANDLE

CANNON LOADING HANDLE

CANNON CHARGING HANDLE

2.1.5-d P-39 Cockpit and Instrument Panel

31

2.1.5-e   Close-up of YP-39 port side of cockpit and instrument panel

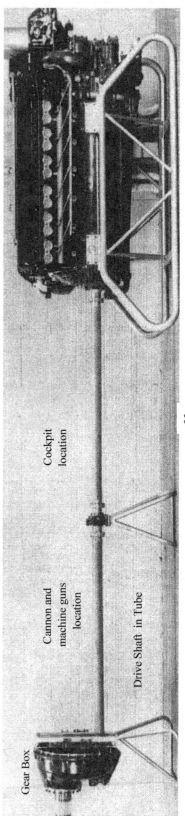

2.1.5-f P-39 Allison V-12 engine and drive train

Cockpit
location

Cannon and
machine guns
location

Drive Shaft in Tube

Gear Box

2.1.5-g   P-39 production line at Wheatfield Plant

34

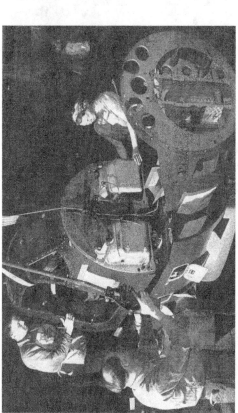

2.1.5-h   Assembling the P-39 at the Wheatfield Plant

35

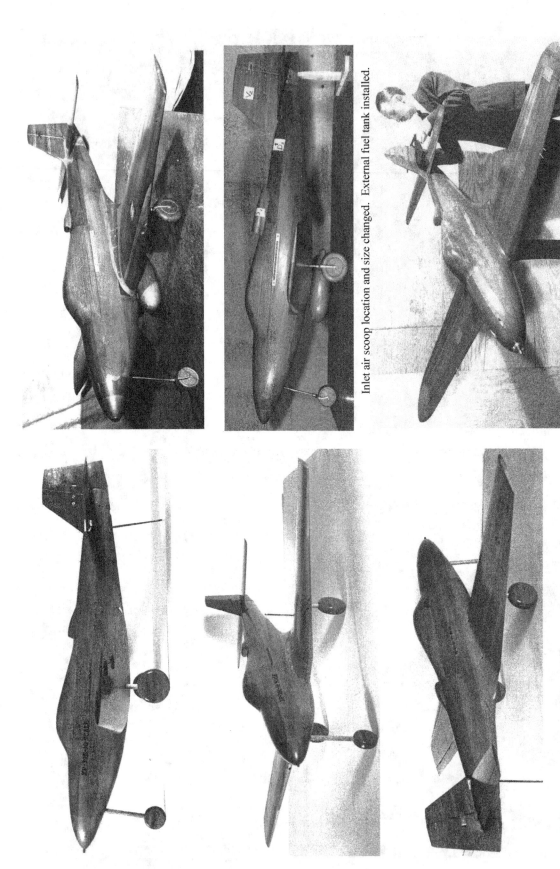

2.1.5-i Wooden P-39 scale models used for wind tunnel testing. Wheels removed when testing in wind tunnel.

Inlet air scoop location and size changed. External fuel tank installed.

Preparing model for wind tunnel

Tail section design changed on1/12 scale model.

36

2.1.6 WWII Bell P-39Q-15BE-44-2911 Airacobra Recovered From Russian Lake

Plane pulled out by truck using a cable

Starboard wing was damaged during earlier combat mission; repairs were made.

Fabric covered control surfaces degraded with time.

Ammunition compartment cover still attached to wing.

Tip of port wing still attached when plane was pulled from lake. Earlier, tip was never painted, after retrofit.

Photos by Boris Osetinskij

Flotation balloons to support nose, during recovery

Both car-like cockpit doors were closed. Pilots' remains were found inside. The safety harness was unlatched, suggesting the pilot may have died on impact.

Russian insignia partially covered by reinforcement plate.

2.1.6-a Bell P-39Q, on Lend-Lease to Russia during WW II, being recovered from bottom of Russian lake, after 60 years

37

Flotation balloons to support nose, during recovery.

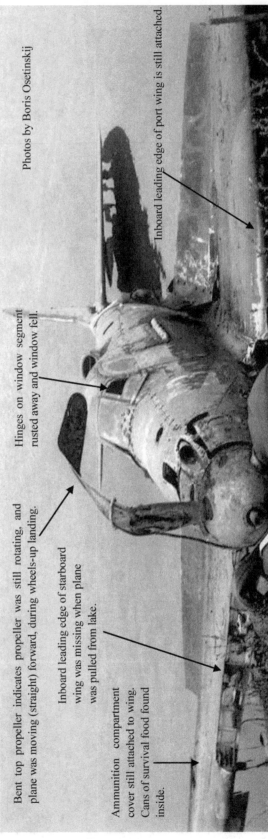

Photos by Boris Osetinskij

Bent top propeller indicates propeller was still rotating, and plane was moving (straight) forward, during wheels-up landing.

Inboard leading edge of starboard wing was missing when plane was pulled from lake.

Hinges on window segment rusted away and window fell.

Ammunition compartment cover still attached to wing. Cans of survival food found inside.

Inboard leading edge of port wing is still attached.

2.1.6-a Bell P-39Q, on Lend-Lease to Russia during WW II, being recovered from bottom of Russian lake after 60 years

38

Russian P-39Q Being Taken Apart at Ira G. Ross Museum

Compartment for:
2 – .50 cal. machine guns/ammunition
1-cannon/ammunition

Engine compartment

New four blade propeller

Original three blade propeller cut short by salvage team

Nose cannon

Carriage

Cenkner Photo

2.1.6-c   The fuselage is mounted on a carriage that was built by the salvage team

As part of the WW-II Lend-Lease program, P-39's were being flown from the U.S. to Russia, through Siberia. These planes were to be used on the eastern front, where they proved to be outstanding ground attack support aircraft and formidable "tank busters', due to their unique nose cannon. They probably helped win the war because they helped prevent Russia from being overrun at the beginning of the war, since Russia didn't have a credible air force, at that time, due to a lack of airplanes; some of them were wooden. One of these P-39's was forced to make an emergency landing on a frozen Siberian lake; the pilot didn't survive. When the ice cracked on impact or melted, it sank to the bottom. It laid there for 60⁺ years, until it was reported by a local fisherman in 2004.

The Russian's were in the habit of storing survival food and clothing in both wings. When the plane was recovered, the survival gear was found -- still in both wings. It was also observed that the leading edge of one of the wings was missing. At first, it appeared that the plane was forced down because the leading edge wasn't properly installed, after the survival gear was inserted; closer investigation revealed that this was not the case.   When the plane was recovered, the decorated pilot's body was found, still in the cockpit. He was given a hero's burial in Russia.

The Ira G. Ross Museum purchased the recovered plane from the salvage company, so that it could be returned to the Bell plant, where it was built. It is believed that this is the first time that a WW-II plane has ever been returned to its production plant.

2.1.6-b   P-39 starboard wing, at lake, after removal from craft.

The starboard wing was removed from the craft after the plane was recovered from the lake. This photo shows that all the top wing covers are missing. However, Figure s 2.1.6-a clearly reveals that all these covers were installed when the plane was pulled from the lake. The missing covers, in this photo, were therefore taken off by the recovery crew, after the plane was recovered; they could not have played any part in bringing down the plane.

There is no photographic evidence that shows that the ammunition cover, on the bottom of the wing, was still attached when the plane was removed from the lake. The fact that the top wing covers were remover at the recovery site, along with numerous other covers, strongly suggests that this bottom cover was also removed there.

All covers on top of wing removed

Inboard leading edge missing

Photos by Boris Osetinskij

Russian P-39Q Being Taken Apart at Ira G. Ross Museum

Compartment for:
2 - .50 cal machine guns/ammunition
1-cannon/ammunition

Engine compartment

New four blade
propeller

Carriage

Cenkner Photo

Nose cannon

Original three blade
propeller cut short by
salvage team

2.1.6-c The fuselage is mounted on a carriage that was built by the salvage team

As part of the WW-II Lend-Lease program, P-39's were being flown from the U.S. to Russia, through Siberia. These planes were to be used on the eastern front, where they proved to be outstanding ground attack support aircraft and formidable "tank busters", due to their unique nose cannon. They probably helped win the war because they helped prevent Russia from being overrun at the beginning of the war, since Russia didn't have a credible air force, at that time, due to a lack of airplanes; some of them were wooden. One of these P-39's was forced to make an emergency landing on a frozen Siberian lake; the pilot didn't survive. When the ice cracked on impact or melted, it sank to the bottom. It laid there for 60+ years, until it was reported by a local fisherman in 2004.

The Russian's were in the habit of storing survival food and clothing in both wings. When the plane was recovered, the survival gear was found -- still in both wings. It was also observed that the leading edge of one of the wings was missing. At first, it appeared that the plane was forced down because the leading edge wasn't properly installed, after the survival gear was inserted; closer investigation revealed that this was not the case. When the plane was recovered, the decorated pilot's body was found, still in the cockpit. He was given a hero's burial in Russia.

The Ira G. Ross Museum purchased the recovered plane from the salvage company, so that it could be returned to the Bell plant, where it was built. It is believed that this is the first time that a WW-II plane has ever been returned to its production plant.

40

2.1.6-b P-39 starboard wing, at lake, after removal from craft.

The starboard wing was removed from the craft after the plane was recovered from the lake. This photo shows that all the top wing covers are missing. However, Figures 2.1.6-a clearly reveals that all these covers were installed when the plane was pulled from the lake. The missing covers, in this photo, were therefore taken off by the recovery crew, after the plane was recovered; they could not have played any part in bringing down the plane.

All covers on top of wing removed

Inboard leading edge missing

Photos by Boris Osetinskij

There is no photographic evidence that shows that the ammunition cover, on the bottom of the wing, was still attached when the plane was removed from the lake. The fact that the top wing covers were remover at the recovery site, along with numerous other covers, strongly suggests that this bottom cover was also removed there.

Ammunition compartment cover on bottom of wing removed

39

2.1.6-d Front side of both wings of the recovered P-39Q

Notice that the inboard leading edge of the front (right) wing is missing, suggesting that it fell off because it wasn't properly reinstalled after survival gear was inserted in the wing.

The back wing still has the entire leading edge intact, but, its' outboard tip is missing.

Movies taken by the Russian/English salvage team, as the plane was pulled out of the lake, show that the leading edge of the right wing was missing and the edge of the left wing was still attached.

The movie also shows the port wing tip still attached when the plane was pulled out.

2.1.6-e Back side of wing and fuselage

The three blades of the original propeller were cut off in Russia, by the salvage team. The metal carriage frame, supporting the fuselage, was built in Russia by the salvage team. The location of the cannon, in the center of the propeller hub, is just visible .

The ammunition cover, on the bottom of the wing, was shown missing after the wing was removed from the plane, at the salvage site. No explanation was given. However, covers on the topside of the wing were shown removed. Other photos show they were still attached when the plane was removed from the lake. This seems to indicate that the top and bottom wing covers were removed by the salvage crew.

Pictures from salvage site show that all covers on top wing surface were removed at salvage site , after the plane was pulled from the lake

Top wing covers missing. Recovery pictures show the covers were removed at the recovery site.

Port wing tip missing

Starboard wing

Inboard leading edge missing from starboard wing

Three propeller blade tips, bent during the forced landing, were cut off by the salvage team

Bottom surface ammunition cover was shown missing, at salvage site, after wing was removed. No explanation was given.

Starboard wing

Cenkner Photos

41

2.1.6-f Removed P-39Q engine

The engine shows structural damage and the oil pan has completely corroded away, as were some other parts. The two holes in the block (one side only) are each in line with a crankshaft throw.

See discussion in Section 2.2.6-l.

2 holes in engine block

Oil pan corroded away

2.1.6-g Some parts removed from the P-39Q

Most parts are intact, but some have corroded away, including the foreground blower housing.

Cenkner Photos

42

2.1.6-i   P-39Q cockpit with instrumentation panel removed

Pipe covering engine drive shaft.

2.1.6-h   Russian emblem painted over U.S.A. emblem during Lend-Lease Program

Cenkner Photos

43

2.1.6-j   Close-ups of internal view of engine block and thrown connecting rod caps

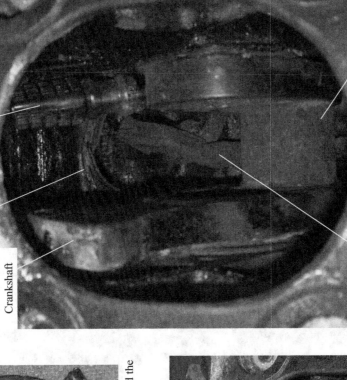

Screwdriver head

Piston pulled down
past bottom of cylinder

Crankshaft

Bent connecting rod

Crankshaft throw. Two piston connecting rods
were attached to each throw, one from each side
of the V-12 engine.

Cenkner Photos

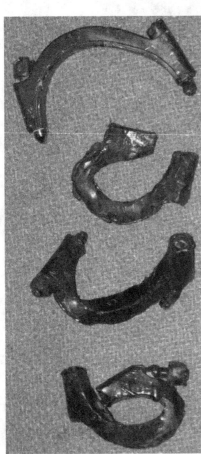

Four connecting rod caps found in the bottom of the aircraft. They were used to attached the
connecting rods to crankshaft throw. The weakest part, the bolts, failed.

44

2.1.6-k Close-ups of region of missing inboard starboard wing section

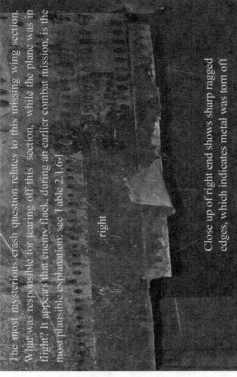

The most mysterious crash question relates to this missing wing section.
What was responsible for tearing off this section, while the plane was in
flight? It appears that enemy flack, during an earlier combat mission, is the
most plausible explanation: see Table 2.1.6-l

right

Close up of right end shows sharp ragged
edges, which indicates metal was torn off

middle

Close up of middle section shows sharp ragged
edges, which indicates metal was torn off

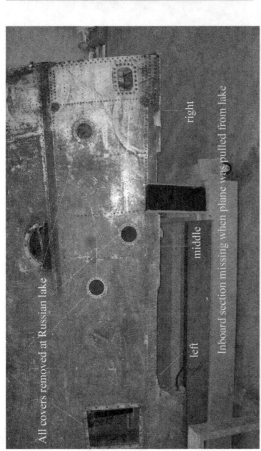

right

left

middle

All covers removed at Russian lake

Inboard section missing when plane was pulled from lake

left

Close up of left end shows sharp ragged
edges, which indicates metal was torn off

Cenkner Photos

45

Issues That Have to be Considered When Identifying the Cause of the Forced Landing of P39Q-15BE-44-2911 Airacobra

The previous images, taken at the recovery site and at the museum, raise a number of issues that must be addresses if a plausible explanation -- for the crash -- is to be identified.

(1) One of the most perplexing unknowns is related to the missing inboard section of the starboard wing (Figure 2.1.6-k). In particular when, how, and why was it torn off the wing?

(2) Why and when were four connecting rod caps torn off their connecting rods and thrown through the side of the engine block (Figure 2.1.6-f & 2.1.6-j)?

(3) How was a piston pulled below the bottom of its cylinder wall (Figure 2.1.6-j)?

(4) What happened to the top wing covers on the port and starboard wings (Figures 2.1.6-d)?

(5) What happened to the bottom wing covers on the port wing (2.1.6-e)?

(6) All three propeller blades were bent. Why was the top propeller blade bent (Figure 2.1.6-a)? This shows that the engine rotated at least one full revolution after the plane landed on the ice.

(7) Why didn't the pilot exit the plane after the crash?

(8) Why did the pilot suddenly yell "port" and then dive out of the formation? This seems to suggest there was a sudden unexpected catastrophic event.

2.1.6-1   What forced the Russian pilot to leave formation and crash land, only 29 km from his destination airfield?

A study was conducted to identify the most probable cause of the forced landing of Bell's P-39Q-15BE-44-2911 Airocobra aircraft, which was part of the Russian Lend-Lease program. The study included inspecting the disassembled P-39Q at the Ira G. Ross Aerospace Museum, in addition to reviewing a translation of the planes maintenance record, two cited articles, and examining pictures taken during the recovery from the Russian lake Mart-Yavr, which is 29km SE of the destination airfield at Luostari.

On November 19, 1944, the day of the crash, the plane was flying in a formation with other Russian aircraft, on a 65 mile non-combat mission to ferry the aircraft to the airfield at Luostari. The plane was operated by an experienced Russian combat pilot who had flown 90 missions with his unit; seven enemy planes had been shot down on previous combat missions. The plane had been hit by enemy flak, during an earlier combat mission, that damaged the starboard wing fuel tank and filler tube; apparently no other damage was detected or reported.

While flying in formation the pilot suddenly shouted port, turned 90 degrees and dropped out of the formation; his wing man initially followed him, but he returned to formation when ordered to do so. The pilot was never heard from again. His body was discovered in the cockpit when the plane was pulled from the lake on August 19, 2004.

Three theories were initially proposed (Theories 2,3,4) to explain the forced landing. As I studied the wreckage at the museum, in addition to video and still photographs of the recovery, it struck me that there is another more plausible explanation—one that seems to explain all of the observed crash phenomena (Theory 1). The enclosed table summarizes these four theories and the corresponding conclusions; this is followed by a detailed discussion of the table and what was uncovered during the study. The conclusions are backed up by numerous included photographs that were taken at the recovery site and at The Museum.

PROPOSED THEORY	EVALUATION OF THEORY	COMMENTS
1) Inboard leading edge of starboard wing was torn off, while in flight, by aerodynamic forces. This probably resulted from undetected wing damage caused by flak, during an earlier combat mission. The pilot lost effective control of the plane due to the disruption of airflow over the wing.	Most probable.	This theory appears to explain all facts that are available at this time.
2) The engine froze up in flight -- throwing four connecting rod caps thru the side of the engine -- due to extremely dirty oil; the Russian policy was apparently to only refill the oil when low but not replace it. Presumably this was due to a wartime shortage of oil.	Improbable.	Does not appear to explain all observations. e. g.: (1) Why was the top propeller blade bent if the engine froze up.(Figure 2.1.6-a) (2) What caused the inboard leading edge of the starboard wing to be torn off; photographs show it was missing when the plane was pulled from the lake (Figure 2.1.6-a).
3) The plane ran out of fuel.	Highly improbable.	The plane was flying in formation with other similar planes that did not run out of fuel. There should have been enough time for the pilot to alert the squadron leader that he was running low on fuel. A sudden unexpected event appears to have forced him to leave the formation and lose altitude.
4) The cover on the starboard ammunition compartment wasn't properly secured when survival food was placed in the compartment. The cover came off in flight and disrupted the air flow over the wing. This caused the pilot to lose effective control of the craft.	Impossible.	Ammunition compartment covers on both wings, along with all other wing covers, were still attached to the wings when the plane was recovered from the lake.

2.1.6-1   Evaluation of Proposed Theories to Explain Crash Landing of Russian P-39   Cenkner Table

THEORY(1): INBOARD LEADING EDGE OF STARBOARD WING WAS TORN OFF DURIND FLIGHT – MOST PROBABLE CAUSE

An overview of the series of events, leading up to the loss of the plane, is first given and then each event is discussed in more detail.

According to this theory, the basic sequence of events leading up to the loss of the P-39Q, with some variations, would be.:

(i)   During an earlier mission, sections of the starboard wing were damaged by enemy artillery flak. Mechanics didn't detect the damage or considered it insignificant.

(ii)   Repeated flexing of the wing resulted in cracks being formed, in the inner leading edge of the starboard wing, that eventually spread and resulted in the section being torn off by aerodynamic forces

(iii)   The pilot lost complete control of the craft and was forced to make a landing.

(iv)   The engine was still operating when the pilot landed, wheels up, on a frozen lake.

(v)   The pilot was flying with his safety harness disconnected. He didn't have time to connect it before landing, so he probably died upon impact.

(vi)   When the plane was forced to land wheels up, on a frozen lake, the propeller was bent at impact. This shows that the engine was forced to stop abruptly, which threw the connecting rod caps through the side of the engine block.

(vii) The rotating propeller cracked the thin ice. The plane sank before the pilot had a chance to escape, if he survived the crash.

DISCUSSION OF ABOVE FACTORS

(i)   During an earlier mission, sections of the starboard wing were damages by enemy artillery flak. Mechanics didn't detect the damage or considered it insignificant.

Discussion of (i)

It was reported that the starboard wing had been hit by enemy flack -- during a previous combat mission -- and this caused damage to the starboard fuel tank and filler tube. The damage was repaired the same day and the plane returned to action the next day. No additional information has been found about this incident. Since the fuel tank is installed inside the wing, this seems to suggest that shrapnel penetrated the wing. With the wing fuel tanks in close proximity to the missing inboard leading edge. this also seems to imply that the inboard section of the leading edge of the starboard wing was also sprayed with metal shrapnel.

Shrapnel doesn't necessarily bring down a plane, unless it hits a critical structural area , equipment or crew members. Unfortunately, from an engineering point of view, the inboard leading edge turns out to be a critical aerodynamic component. In their rush to return the plane to combat,, maintenance personnel may not have noticed this shrapnel penetration or they may have considered it insignificant.

(ii)   Repeated flexing of the wing resulted in cracks being formed, in the inboard leading edge of the starboard wing, that eventually spread and resulted in the section being torn off by aerodynamic forces

Discussion of (ii)

If the inboard leading edge of the starboard wing was perforated by flying (flack) shrapnel, then this could account for the eventual loss of the plane. Holes in the material would act as stress concentrators, significantly increasing the tensile stress in the wing over what would be obtained if there were no perforations. The amount of stress increase would depend upon the size, shape, number and distribution of the holes – but it would be significant.

When aerodynamic air pressure is applied to the bottom of the wing, during flight, the wing will curve upward, relative to the fixed end at the fuselage. This puts a tensile stress in the bottom fibers of the wing and a compressive stress in the upper fibers. The maximum stretching and stress would occur near the fuselage – in the region where the wing section is observed to be missing. As mentioned above, this maximum stress would be increased substantially if flack had penetrated this section of the wing.

It is known that the material will tear if a certain critical stress is exceeded or it will fail in fatigue, at a lower stress level, if a certain number of pressure loading cycles are applied. If either of these situations occurred, the material would separate and cracks would be formed; the original curvature of the leading edge would probably aggravate the situation. Once the material separates, aerodynamic forces would complete the process of tearing the exposed material from the wing.

Close-up photographs of this inboard region (Figure 2.1.6-k), reveals remaining sharp material segments that are sometimes straight and sometimes irregular, which indicates that the material was torn off.

The shape of the leading edge, material properties, and engine vibrations probably also played a part in the failure of this part.

Discussion of (iii)

(i)   The pilot lost complete control of the craft and was forced to make a landing.

Once the inboard leading edge was torn off, the air flow over the wing was changed dramatically. The lift would decrease and the plane probably became unstable.

There was one radio transmission from the pilot, issuing the command "port" to his wing man, then he lost altitude. This suggests it was a sudden unexpected event that forced him to leave formation and make a crash landing. A violent tearing off of a section of the wing would certainly be such an event.

Discussion of (iv)

(ii)   The engine was still operating when the pilot landed, wheels up, on a frozen lake.

In Figure 2.1.6-a the vertical blade of the propeller is bent, indicating that the propeller and engine were still turning, at landing, and that the wheels were up. When the blades hit the ice, they were bent and they probably cracked the thin ice.

Discussion of (v)

(iii)   The pilot was flying with his safety harness disconnected. He didn't have time to connect it before landing, so he probably died on impact.

The pilot was found in the cockpit, with both doors closed, and his safety harness unbuckled.. With a forced landing, the pilot should have jettisoned at least one door. In flight, Russian pilots would disconnect their safety harness to give themselves more freedom, as they searched the skies for enemy airplanes – one explanation for the disconnected harness. Another explanation would be that he disconnected the harness after he landed. With an experienced fighter pilot, it seems logical that he would have made sure that his harness was connected prior to landing – if he had time.

If the harness was disconnected on landing, he may have perished then. However, if the harness was connected at landing then he disconnected the harness, bur did not have time to exit the plane before it fell through the ice. The bent propeller suggests there may have been enough force to crack the thin ice at impact.

(i) When the plane was forced to land, wheels up, on a frozen lake, the propeller blades were bent at impact. This shows that the engine was forced to stop abruptly, which threw the connecting rod caps through the side of the engine block.

Discussion of (vi)

This abrupt forced stopping of the engine crankshaft, while the pistons were still firing and producing power, twisted the connecting rods (Figure 2.1.6-j), pulled four connecting rod caps (Figure 2.1.6-j) off the crankshaft throws and sent them through the side of the engine block (Figure 2.1.6-f).

(ii) The rotating propeller cracked the thin ice. The plane sank before the pilot had a chance to escape, if he survived the crash.

Discussion of (vii)

Since the plane landed with the wheels up, the rotating propeller made contact with the ice and was bent (Figure 2.1.6-a). It was reported that the ice was relatively thin at this time of year, so the ice was probably cracked when the propeller impacted. With the plane being found still inside the craft, he was either killed on impact or the plane sank so fast he didn't have a chance to exit the cockpit This is supported by the fact that both cockpit doors were closed, when the plane was found

The pilot was found in the cockpit, with his safety harness unbuckled. Either: (1) His harness was unbuckled when he landed, which probably means that he didn't survive the landing (Russian pilots flew combat mission with unbuckled harnesses, so they could better scan the sky foe enemy aircraft) or (2) he unbuckled his harness upon landing but didn't have time to exit..It would seem that his harness was probably unbuckled; it's doubtful he had time to buckle it, once the plane left formation. What actually happened here has no bearing on this theory.

THEORY(2): ENGINE FROZE UP IN FLIGHT – IMPROBABLE

According to this theory, the sequence is:

(i)   The engine suddenly froze up in flight, due to sludge in the oil.
(ii) The pilot was forced to suddenly leave formation and land.

DISCUSSION OF ABOVE FACTORS

(i)   The engine suddenly froze up in flight, due to sludge in the oil.

Here it's postulated that oil sludge plugged up some of the engine block oil feed holes. With no lubrication, metal piston against metal cylinder wall resulted in some pistons freezing up. Concurrently, other pistons were still moving and providing power to the crankshaft. With some pistons frozen in place and the crankshaft being rotated by other pistons, the frozen pistons connecting rods were bent and twisted, while the weakest link -- the connecting rod cap bolts-- failed. The caps were thrown off the crankshaft and through the side of the cylinder block.

The sludge buildup occurred because the Russians frequently didn't change engine oil; they merely replenished it. This was probably due to a wartime shortage of oil.

While this explanation seems logical, at first blush, it doesn't explain other observed crash phenomena:

(1) The top propeller blade was bent during the landing; Figure 2.1.6-a. The propeller had to be turning for the blade to end up in the vertical position. If the engine was frozen, this obviously couldn't have happened.

50

THEORY (3): THE PLANE RAN OUT OF FUEL – HIGHLY IMPROBABLE

With this theory, the sequence is:

(i)   The plane ran out of fuel.
(ii)   The pilot was forced to suddenly leave formation and land.

DISCUSSION OF ABOVE FACTORS

(i)   The plane ran out of fuel.

Discussion of (i)

The plane was flying in formation, with other Russian planes, to a new Russian air field. During the incident, the only communication from the pilot was the command "port" to his wingman, indicating he was turning to the left. He then turned 90 ° - 120 ° to the left and he lost altitude. This certainly suggests there was a sudden unexpected event that forced him to execute this maneuver – running out of gas doesn't seem to fit. If he was running out of fuel, it seems like he would have had plenty of time to report his situation – which he didn't.

While there may have been extenuating circumstances, it seems logical that all of the planes in formation would have been refueled before the flight. No report was received that other planes ran out of fuel.

As in theory 2, this theory fails to explain other observed crash phenomena: the bent propeller, the missing starboard wing segment, and the thrown piston rod caps.

THEORY (4): AMMUNITION COMPARTMENT COVER WASN'T PROPERLY SECURED WHEN FOOD WAS PLACED INSIDE – IMPOSSIBLE

According to the final theory:

(i)   Ammunition was removed from the starboard wing compartment and emergency food cans were inserted.
(ii)   The starboard wing compartment cover was improperly reinstalled.
(iii)   The cover was torn off in flight, causing the plane to become unstable and forcing a landing.

DISCUSSION OF ABOVE FACTORS

(i)   Ammunition was removed from the compartment and emergency food cans were inserted.

Discussion of (i)

Food cans were found in the ammunition compartment of both wings.

(ii)   The starboard wing compartment cover was improperly installed.

Discussion of (ii)

It's clear, in Figure 2.1.6-a, the ammunition compartment covers on both wings, along with all other covers, were attached to the wing when the plane was pulled from the lake.

51

(iii)   The cover was torn off in flight, causing the plane to become unstable.

Discussion of (iii)

Even if the cover was torn off, it's doubtful that this would have caused the plane to become so unstable that it couldn't be controlled. The open area of the ammunition compartment is quite small compared to the wing area. A recirculation bubble of air would probably have formed in the compartment and the air would have passed over the top.

References

(1) DeCroix, Douglas W, *From Russia ... With Log*, Western New York Heritage, Vol. 12 No. 3, Fall 2009, pg. 32.

(2) Sheppard, Mark, Bell P39Q-15BE-44-2911 Airacobra, http://lend-lease. airforce.ru/english, 17.09.2006.

(3) *HANDBOOK OF ERECTION AND MAINTENANCE INSTRUCTIONS FOR THE P-39 SERIES AIRPLANE*, Bell Aircraft Corporation, Buffalo, N.Y., December 1, 1942.

(4) Ira G. Ross Aerospace Museum, *Bringing Her Home: The Story of the Miss Lend Lease*, DVD, YY-06-57.

(5) Pilkey, Walter D., Peterson's *Stress Concentration Factors*, John Wiley & Sons, Second Edition, 1997.

(6) *PILOTS FLIGHT OPERATING INSTRUCTIONS FOR ARMY MODEL P-39Q-1 AIRPLANE*, Aviation Publications, Appleton, Wisconsin.

Twenty-two years old.

He was an experienced combat pilot, having flown 90 operational missions with his unit.

His remains were found in the cockpit when the plane was recovered.

He was last seen while flying in formation with other Russian planes. They were ferrying the planes to a new base in Russia. He suddenly yelled port and dove to his port side – no other communication was received. His wing man dove with him but he was called back by the formation commander. The Russian military thought he had deserted.

Lt. Baranovsky's family did not know of his recovery until they were notified by the Ira G. Ross Aerospace Museum.

He was buried with full military honors at the Glory Valley Memorial near Murmansk, Russia.

As part of the research on this book, the author conducted an engineering study of the recovered plane, and photographs made during its recovery. It was concluded that the plane was probably brought down by starboard wing flak damage that was sustained during an earlier combat mission. Flak holes, in the inboard leading edge of the starboard wing, would have acted like stress concentrators – greatly increasing the localized stress in the wing. Continued flexing and vibration would have eventually produced cracks in this curved region of the wing. In-flight aerodynamic forces would have resulted in segments of the inboard section of the starboard wing being suddenly torn off. This would have disrupted the air flow over the wing, causing the pilot to suddenly lose control of the plane. When the plane crash landed, with the wheels up, the abrupt stopping of the engine – as the propellers hit the ground -- resulted in two connecting rod caps being torn off and thrown through the side of the engine block. This scenario explains all of the observed plane damage, as well as the abrupt behavior of the pilot. A more detailed discussion is given in the beginning of this section.

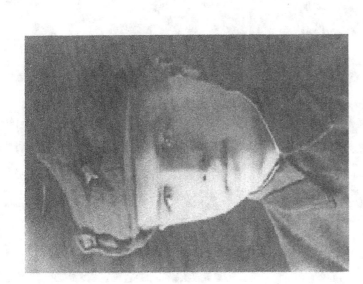

2.1.6-m   Lt. Ivan Ivanovich Baranovsky, Russian P-39 Pilot.

2.1.7 Bell's P-63 Kingcobra Fighter and Ground Attack Airplane

2.1.7-a Bell's P-63 with Russian star on side in preparation for flying to Russia as part of Lend-Lease Program

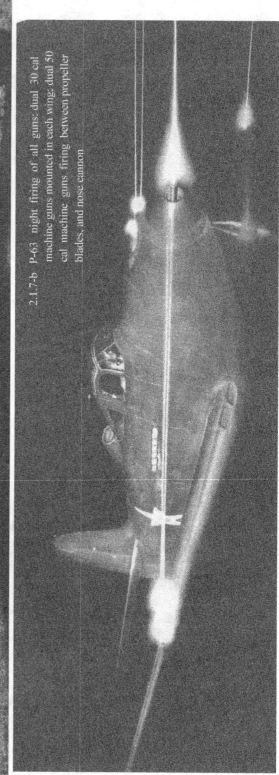

2.1.7-b P-63 night firing of all guns: dual 30 cal machine guns mounted in each wing; dual 50 cal machine guns firing between propeller blades, and nose cannon

53

2.1.7-c   P-63 twin 50 caliber machine guns mounted in fuselage

Side view of twin 50 caliber machine guns, mounted in frame. Also shows 37 mm cannon through center of propeller.

U. S. Army Photo

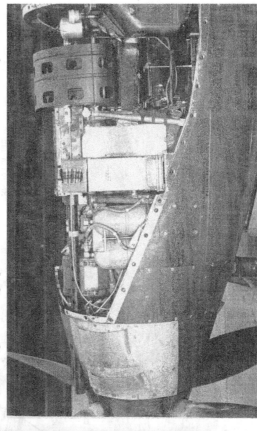

Side view of fuselage, with panels removed, revealing 50 caliber machine guns, ammunition cartridges and ammunition storage boxes

Front view of fuselage with panels removed. Twin 50 caliber machine guns are synchronized with rotating propellers so they can fire between the rotating blades.

54

2.1.7-d   Exploded View of Bell Aircraft P-63 Kingcobra

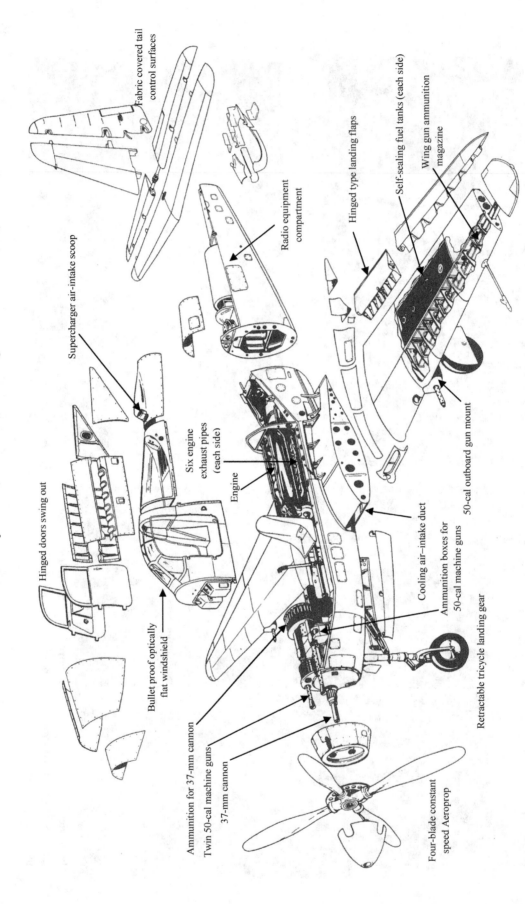

Fabric covered tail control surfaces

Self-sealing fuel tanks (each side)

Wing gun ammunition magazine

Hinged type landing flaps

Radio equipment compartment

Supercharger air-intake scoop

50-cal outboard gun mount

Six engine exhaust pipes (each side)

Engine

Hinged doors swing out

Cooling air-intake duct

Ammunition boxes for 50-cal machine guns

Bullet proof optically flat windshield

Ammunition for 37-mm cannon

Twin 50-cal machine guns

37-mm cannon

Retractable tricycle landing gear

Four-blade constant speed Aeroprop

55

2.1.7-e Bell P-63 Kingcobra Assembled Phantom Drawing

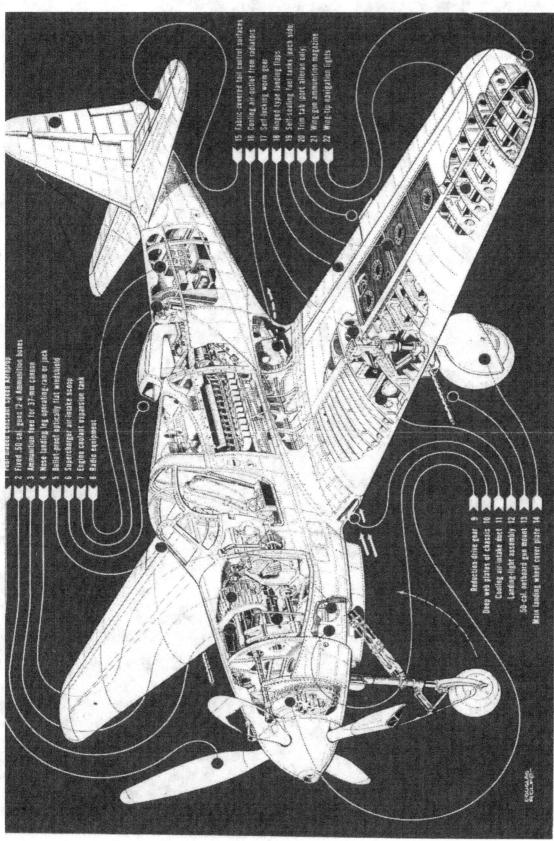

1 Fabricated cockpit space frame
2 Fixed .50-cal. guns (2-4 Ammunition boxes
3 Ammunition lines for 37-mm cannon
4 Nose landing leg operating cam or jack
5 Bullet-proof optically flat windshield
6 Supercharger air-intake scoop
7 Engine coolant expansion tank
8 Radio equipment

9 Reduction drive gear
10 Deep web plates of chassis
11 Cooling air-intake duct
12 Landing-light assembly
13 .50-cal. outboard gun mount
14 Main landing wheel cover plate

15 Fabric-covered tail control surfaces
16 Cooling air-outlet from radiators
17 Self-locking worm gear
18 Hinged-type landing flaps
19 Self-sealing fuel tanks each side
20 Trim tab sport aileron only
21 Wing gun ammunition magazine
22 Wing top navigation lights

Faster and with a higher effective combat altitude and greater range than it's predecessor, the P-39 Airacobra, the P-63 Kingcobra is rated at a speed "well over 400 mph" and is called an able fighter at an altitude up to 40,000 feet. Bigger fuel tanks add range.

56

2.1.7-f Close-up of P-63 port side of cockpit and instrument panel

2.1.8 Bell's All Wooden XP-77 Airplane

2.1.8-a XP-77 largely built of wood due to anticipated shortage of metal during WWII

2.1.8-b Close-up of wooden XP-77 with designer Bob Woods. Mr. Woods left Consolidated Aircraft, with Larry Bell, and worked for Bell Aircraft Corp. as its first airplane designer.

2.1.9 Bell's P-59 Jet Airplane

2.1.9-a Bell's Data Sheet for P-59 Jet Aircraft

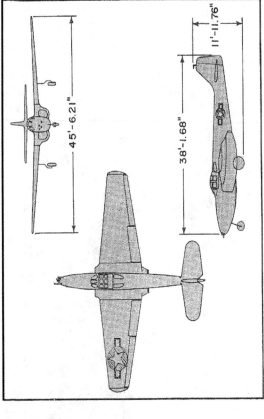

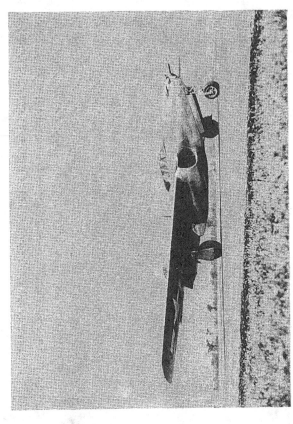

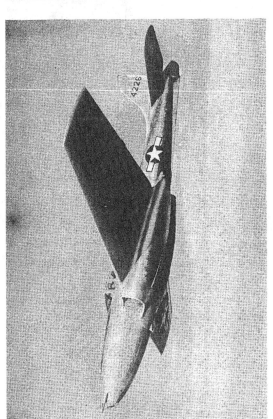

Bell P-59 Airacomet

This was the first United States airplane which discarded conventional internal combustion engine and propellers for the revolutionary jet propulsion engine. Bell was asked three months before Pearl Harbor to design and build a high-performance fighter, powered with General Electric units fashioned after the British Whittle design. The P-59 made its first flight on Oct. 1, 1942—less than 13 months later. Bell built 57 by the end of World War II and the first now is in the Smithsonian Institution. Equipped with a tricycle landing gear and the two GE turbo-jet engines, each capable of 1600 pounds of thrust, the Airacomet had a speed of more than 400 miles an hour and a service ceiling above 40,000 feet. Subsequent models were used by Bell in developing remote radio control.

BELL Aircraft CORP.

BUFFALO, N.Y.   FORT WORTH, TEXAS

59

2.1.9-b Top View of Bell's P-59 Jet Fighter

2.1.9-c A dummy propeller was placed on the first American jet airplane, to mislead observers

2.1.9-d P-59 instrument panel

2.1.10   Bell's XP-83 Long Range Jet Escort Fighter

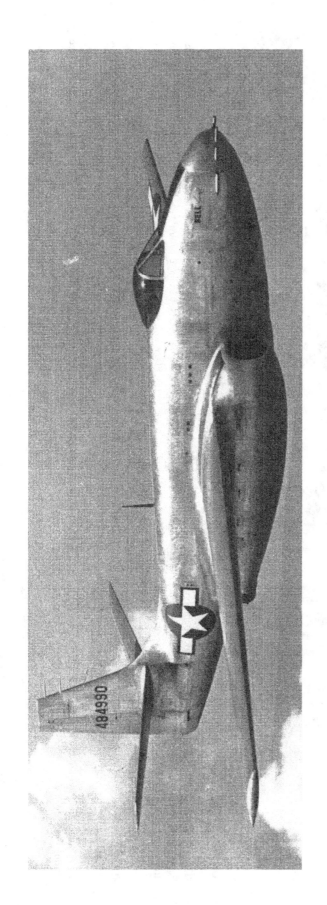

2.1.10-a   The United States Army Air Force contracted Bell to increase the range of jet fighters. Bell installed two General Electric J33-GE-5 turbojet engines, at the base of the wings, to increase the available fuel space in the fuselage; drop tanks were also installed under each wing. Fuel capacity was increased to 1,150 gal, with the drop tanks adding an additional 500 gal, combined. This significantly increased range -- drop tanks installed: 2,050 miles; no drop tanks installed : 1,730 miles.

2.1.11   F-7F Robotic Airplane Retrofitted with Bell's Radio Control System

2.1.11-a   Grumman F7F Tigercat – the first US Navy twin engine fighter -- with Bell's radio control system installed

On 30 June 1941, Grumman awarded a contract to Bell to develop a prototype XF7F-1 robotic airplane.

The goal of this project was to remotely control the operation of a high speed airplane while it was performing various maneuvers: take-off and landing, climbs, right and left turns, level flight, dive to an including 90 degrees from the horizontal, and pull-outs. To achieve this, as shown in Figure 2.1.111-b, remote control was installed for the elevator trim tab, throttle, landing gear flaps, propeller pitch, brakes, etc. Radio telemetry was also installed in the Robot Plane to transmit dynamic flight data and instrumentation data to the Ground Station. Televised images of the flight instruments were required for remote control during some maneuvers.

The test phase required three interacting components: the Robot Plane (radio controlled F7F-3), the Mother Plane (an F-6F-3 control plane), and the Ground Control Station. The Robot could be remotely controlled by the Mother Plane or by the Ground Control Station. Three Bell Aircraft pilots were at the controls of the three stations.

After some structural modifications the Robot Plane passed aircraft carrier qualification tests, but most of the craft were land based; only 12 units were built. Some saw service during the Korean war as land-based night fighters or attack aircraft.

Grumman's F7F-4N Tigercat

Maximum speed:	460 mph
Range:	1,000 nm
Service ceiling:	40,000 feet
Power plant:	2x Pratt Whitney R-2800-34W radial engines (2,199 hp each)
Max takeoff weight:	25,720 lb

Reference

(1)   Fabricy, J. Jr., *Radio Control of F7F-3 Airplane*, Bell Aircraft Corporation, 3-15-46.

63

2.1.11-b   Schematic of Bell's radio control system installed in an F-7F robotic airplane

1. TELEMETERING ANTENNA
2. TELEMETERING POWER SUPPLY
3. BALLAST
4. POWER BRAKE UNIT
**5. LANDING GEAR ACTUATOR
6. ROLL GYRO
7. AUTO-PILOT CONTROL HEAD
8. PILOT'S SWITCH PANEL
**9. AUTO PILOT SERVO ADAPTER BOX
10. RADIO CONTROL STICK GRIP
**11. THROTTLE QUADRANT
12. AUTO PILOT TRAY (SWITCHBOARD, SERVO
     AMPLIFIER, PANEL JUNCTION BOX, RELAY
     JUNCTION BOX)
**13. R.P.M. DRIVE MOTOR
**14. THROTTLE MOTOR
15. AILERON SERVO INSTALLATION
16. ELEVATOR SERVO INSTALLATION
17. FLAP ACTUATOR
18. RADAR BEACON
19. TELEMETERING JUNCTION BOX
20. TELEMETERING COMMUTATOR
21. ELEVATOR TRIM TAB ACTUATOR
22. RUDDER SERVO INSTALLATION
23. TELEVISION DYNAMOTOR
24. TELEVISION FILTER JUNCTION BOX
25. AUTO PILOT INVERTER
26. TELEVISION TRANSMITTER
**27. D.C. POWER BOX
**28. CONTROL JUNCTION BOX

**29. BLOWER AND DIVE RECOVERY CONTROL BOX
30. V.G. RECORDER
31. FUEL SELECTOR UNIT
32. REMOTE CONTROL RECEIVER - F.M.
33. TELEMETERING REGULATOR UNIT
**34. REMOTE CONTROL JUNCTION BOX
35. SUPERSONIC SELECTOR
**36. TELEVISION ANTENNA
37. TELEVISION PHOTO PANEL
38. TELEMETERING TRANSMITTER
39. TELEVISION CONVERSION UNIT
40. REMOTE CONTROL ANTENNA
41. TURN GYRO
42. PITCH GYRO
43. SUPERCHARGER ACTUATOR
44. AUTOMATIC OIL SHUTTER REGULATOR
45. HORIZON GYRO
46. INSTRUMENT PANEL
47. 34 AMPERE STORAGE BATTERY
48. 200 AMPERE GENERATOR
49. APN-33 RADAR ANTENNA
50. BEAM TYPE ACCELEROMETERS
51. ANGULAR TYPE ACCELEROMETER
**52. RADAR BEACON DYNAMOTOR

*   SHOWN ON TOP VIEW ONLY
** SHOWN ON SIDE VIEW ONLY

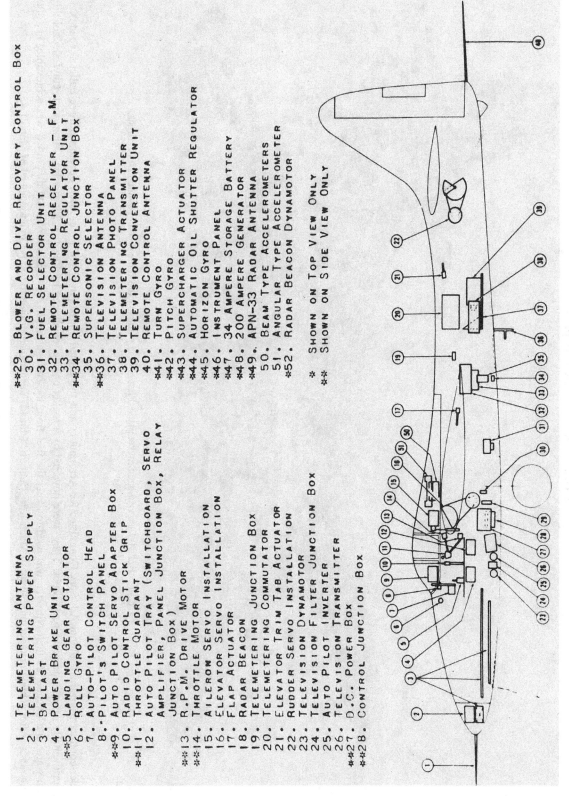

64

2.1.12   Bell X-1Rocket Airplane

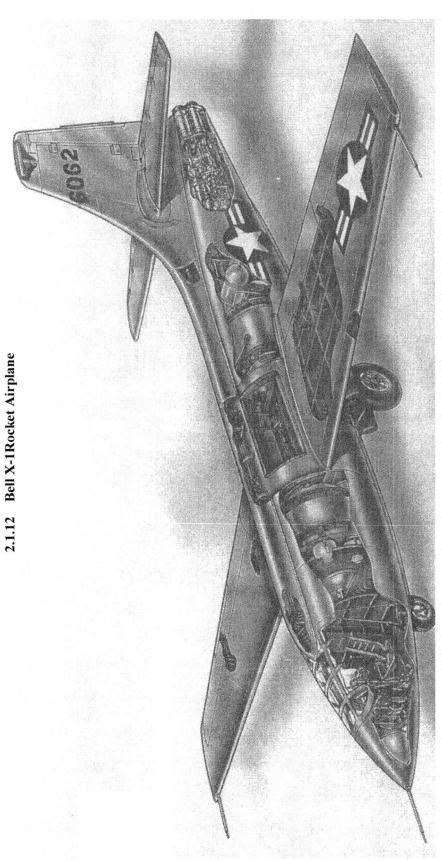

2.1.12-a   Bell X-1 experimental rocket plane. First piloted plane to exceed the speed of sound in level flight.

Under contract from the National Advisory Committee for Aeronautics and the United Stated Army Air Force, Bell developed three bullet shaped X-1s to experimentally investigate high speed flight. They were powered by a rocket engine, developed by Reaction Motors Inc. , that used ethyl alcohol diluted with water and oxygen, as fuel. Nitrogen was used to expel the fuel from the storage tanks. A maximum of 6000 lbf thrust was achievable, in 1500 lbf increments. After achieving supersonic flight at Mach 1.06, Captain "Chuck" Yeager then piloted the X-1-1 to a top speed of Mach 1.45.

65

Pilots cockpit

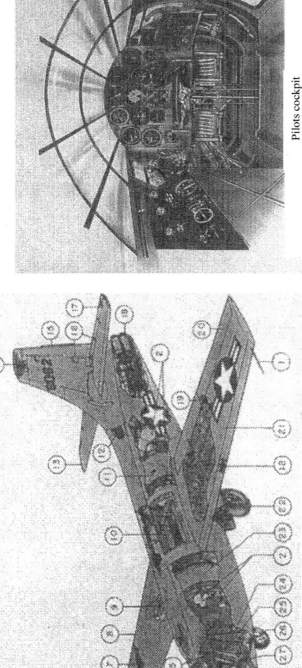

Equipment and controls

1. Pitot Tube
2. High Pressure Nitrogen Tanks
3. Cockpit Entrance R.H. Side
4. Control Stick
5. Head Rest
6. Yaw Angle Vane
7. Airflow Trim Tab Actuator
8. Pilot Controllable Aileron Trim Tab
9. Spoilers
10. Instrument Compartment (Research Equipment)

11. Water Alcohol Tank
12. Pressure Tubes
13. Stabilizer Movement: Up 5^0 Dn 10^0 Pilot Controlled
14. Radio Antenna
15. Rudder: Right 15^0, Left 15^0
16. Rudder Trim Tab
17. Balance Weights
18. Rocket Motor $6000^\#$ Thrust
19. Flaps: 60^0 Movement
20. Ailerons: Travel 12^0 Up 12^0 Dn

21. Tapered Wing Skin
22. Retractable Main Gear
23. Liquid Oxygen Tank
24. Pilots Cockpit (Pressurized)
25. Pilots Shoulder Harness & Safety Belt
26. Retractable Nose Gear
27. Battery
28. Spoilers Control
29. Rudder Pedals
30. Brake Cylinders

2.1.12-b   Bell X-1 equipment, controls, and pilots cockpit.

66

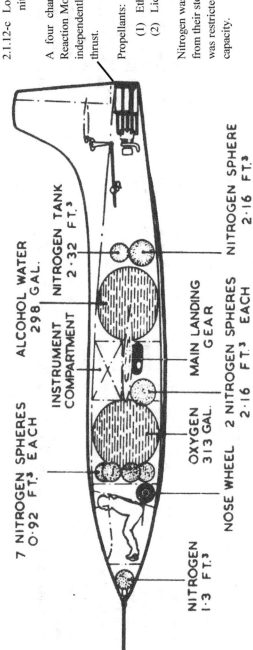

2.1.12-c Location of propellant tanks and nitrogen purge tanks in Bell X-1

A four chamber rocket engine built by Reaction Motors Inc. Each chamber, fired independently, would produce 1500 lbf thrust.

Propellants:

(1) Ethyl Alcohol diluted with water
(2) Liquid Oxygen

Nitrogen was used to expel the propellants from their storage tanks. Full power testing was restricted to 5 minutes because of fuel capacity.

7 NITROGEN SPHERES 0·92 FT.³ EACH

INSTRUMENT COMPARTMENT

ALCOHOL WATER 298 GAL.

NITROGEN TANK 2·32 FT.³

MAIN LANDING GEAR

NITROGEN SPHERE 2·16 FT.³

NITROGEN 1·3 FT.³

OXYGEN 313 GAL.

NOSE WHEEL

2 NITROGEN SPHERES 2·16 FT.³ EACH

2.1.12-d Major components of hydrogen-peroxide rocket engines used in reaction control system for controlling directional stability of the variant X-1B.

After the initial three airplanes were built – X-1-1, X-1-2, X-1-3 -- 5 additional variants were built X-1A, X-1B, X-1C, X-1D, X-1E, to further pursue supersonic development. With slightly different wings, variant X-1B had approximately 300 thermal sensors installed on its surface, for the study of aerodynamic heating. It was flown from October 1954 until it was grounded in January 1955, due to fuel tank cracks.

The wings and control surfaces of the original X-1 proved to be too small for directional control of the craft at high speed. The rocket powered reaction control system was therefore installed to provide the desired control.

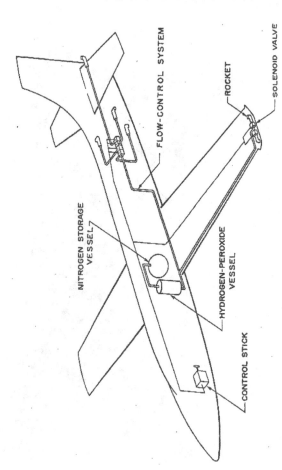

FLOW-CONTROL SYSTEM

NITROGEN STORAGE VESSEL

ROCKET

HYDROGEN-PEROXIDE VESSEL

SOLENOID VALVE

CONTROL STICK

2.1.12-e   X-1 being loaded under the modified B-50 Superfortress mother ship, at Dryden Flight Research Center. Once the mother ship reaches the 30,000 feet release altitude, the X-1 is dropped. While falling, the engine is started and the test flight begins. This saves fuel and extends the available X-1 test time, at full throttle, to about 5 minutes.
NASA Photograph

2.1.13 Bell's X-2 Supersonic Airplane

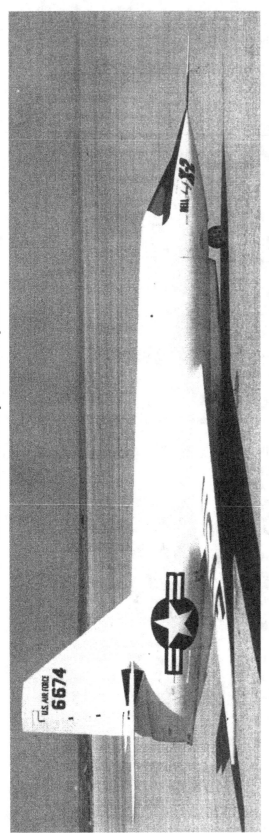

2.1.13-a Side view of Bell's X-2 experimental supersonic airplane

2.1.13-b Front view of Bell's X-2 experimental supersonic airplane

2.1.13-c Bell's X-2 aircraft being mated with the B-50 Mother Ship

2.1.13-d Bell's X-2 aircraft, mated to the B-50 mother ship, in route to the drop zone

71

2.1.14   Bell's X-5 Variable Sweep Wing Jet Airplane

2.1.14-a   Bell's Data Sheet for X-5 variable sweep wing jet airplane

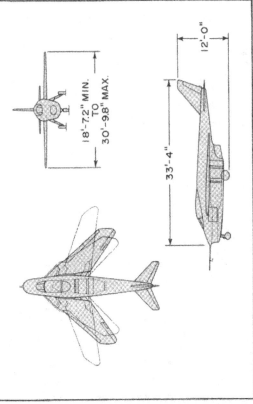

Bell X-5

This is the world's first airplane capable of varying the degree of wing sweepback in flight. For takeoffs and landings, the wings are swept forward to a 20-degree angle. In flight, the wings can be moved to any position between 20 and 60 degrees. Unlike the rocket-powered X-1 and X-1A also built by Bell, the white, needle-nosed craft is jet-propelled and has produced valuable information concerning the aerodynamic effects of sweptback wings. The "flying guppy" configuration results from mounting the Allison turbo-jet engine, which delivers 4900 pounds of thrust, under the cock-pit rather than behind the pilot. The X-5 weighs about 10,000 pounds. Its slender spear-like pitot tube, extending an additional eight feet beyond the nose, also contains yaw-measuring equip-ment. Cockpit canopy and seat are jettisonable.

BUFFALO, N.Y.   **BELL** *Aircraft*   FORT WORTH, TEXAS
CORP.

2.1.14-b   Close up and phantom view of Bell X-5 Jet airplane with swept wings

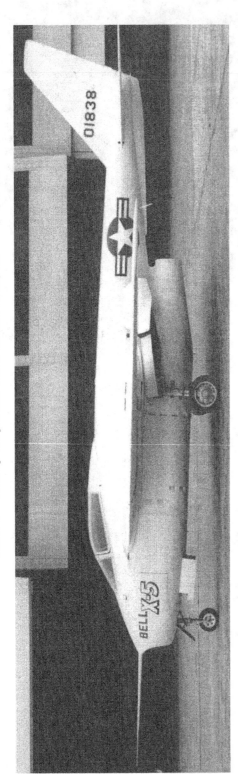

First airplane to change wing sweep angle in flight. The wings could be changed between 20°, 40°, and 60° using a jackscrew assembly.

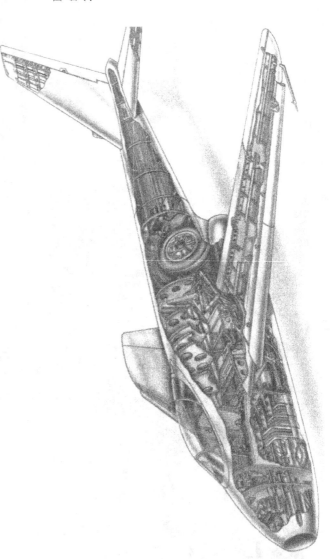

73

2.1.15 B-29 Superfortress Bombers Built By Bell Aircraft

2.1.15-a   The first of 668 B-29s built by the Bell Aircraft Company at its Marietta Georgia plant

The B-29 "Superfortress" long range bomber was developed by Boeing; its first flight occurred on September 21, 1942. To meet government production goals, four assembly plants were required. The two Boeing plants produced 2,766 B-29s, the Bell plant 668, and the Glen L. Martin plant 536.

Marietta, Georgia, a small town North of Atlanta, was selected for the construction of a new production facility that would be operated by the Bell Aircraft Company. The site was selected because there was an existing air field, there was a large labor supply that could be trained in local schools, and nearby companies could produce subcontracting work. The new and only assembly building, which was long enough to enclose more than six in-line football fields, was 2000 feet long x 1000 feet wide, and stood four and a half stories high. An office building and several smaller buildings, were separated from the assembly building. While the production building was windowless, because of the war, it did have air conditioning.

It turned out that the runways were unacceptable: they were too short, too narrow, two thin, and they were pointed in the wrong direction. The U.S. Corps of Engineers ended up spending $2 million on upgrading the runways.

On March 15, 1943 the plant was finished and available for use. Before the last day of 1943, the first two B-29s rolled off the production line. By the end of the war sixty planes per month were being built.

Employment peaked at 28,158 workers with over a third of them women, while 1757 were handicapped, and 1785 were nonwhite. Nine-tenths of Bell employees were from the South.

For supervisory jobs, salaries began at $2,600 a year. Skilled workers received eighty-five cents an hour and unskilled seventy cents an hour. Locals saw their salaries increase by a factor or three or four, over what they earned before joining Bell.

The government offered the production plants a flat-price-per-aircraft contract instead of the standard cost-plus-fixed-fee contract. Only Bell took the offer. Bell Aircraft ended up earning 9.9 percent profit while the other plants only earned 2 percent.

74

2.1.15-b   1000 foot width view -- foreground: preparation of the engine shrouds for installation; background: B-29s being assembled

75

2.1.15-c   2000 foot length view of assembly building showing production area and two floors of office space (left)

76

2.1.15-e   Center fuselage section goes from vertical to final horizontal position

2.1.15-d   Installation of wiring in cockpit

2.1.15-g   Cabin assemblies grow in great two story fixtures

2.1.15-f   Final installation of engines

77

2.1.15-h   Cadmium plating and copper dipping served special purposes

2.1.15-j   Installation of cockpit windows

2.1.15-i   Engine buildup – one of the "pressure spots"

78

2.1.16 Bell's Radio Command Control System for MQM-57A Surveillance Drone

U.S. ARMY

SURVEILLANCE

DRONE

Top Speed: 184 mph
Ceiling: 15,000 feet
Endurance: 40 min.

2.1.16-b   The MQM-57A drone being worked on by Bell engineering aide Robert T. Raseler.

The MQM-57A surveillance drone was developed by the Northrop Corporation; it was operational for 3 ½ years before Bell was contracted to design and build an upgraded radio command control system.

The drone carries a photographic camera, in addition to other surveillance equipment. After the film is exposed, the drone is flown to a recovery area, where the engine is shut off and a parachute is deployed for recovery. The film can then be processed and made available in a matter of minutes.

Bell's new radio command system can transmit commands over longer distances, which greatly increases the operational range of the drone. Bell also installed selectable coding, which can be used to prevent interference with other drones operating in the same area.

The upgrade included specialized equipment that was installed in the drone:
(1) an antenna and (2) electronics required to translate the received signals and actuate the drone's controls.

Ground support equipment can check out the drone flight equipment prior to a mission.

Late in 1963, the first operational system was installed in the field; several hundred would eventually be installed.

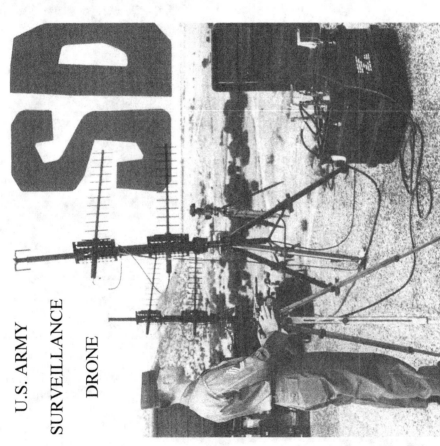

2.1.16-a   The tripod mounted flight control panel, built by Bell, is operated by SFC John F. Brown, a Combat Surveillance School Instructor. In the background are the radio transmitter and the yagi antenna that are used to transmit his commands to the drone. The flight control panel is employed in the navigation of the drone, and to activate lights and the camera, and to deploy the recovery parachute.

Reference

(1) Spindler, Albert W., A "*Peeping Tom*" for the Battlefield, Unknown source.

2.1.17 Bell's Airplane Proposal for a Twin Engine Bomber

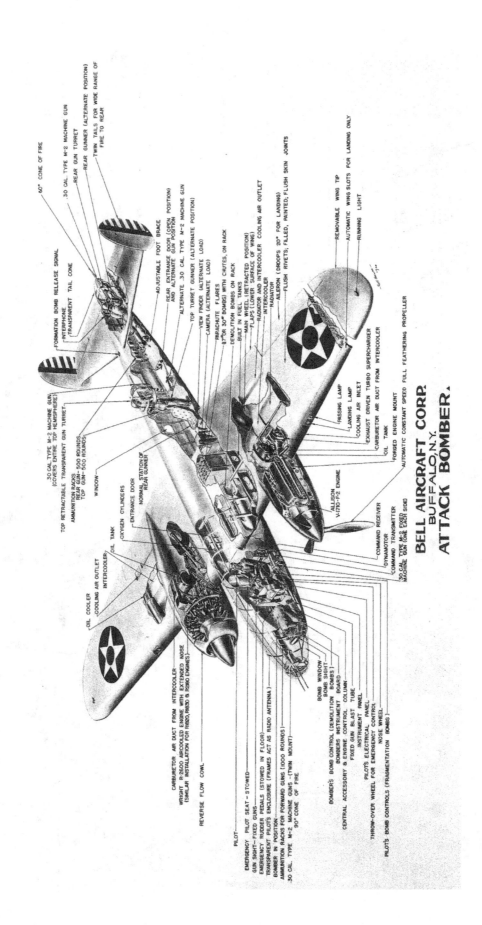

BELL AIRCRAFT CORP.
BUFFALO, N.Y.

ATTACK BOMBER.

2.1.18 Bell's X-16 Reconnaissance Airplane

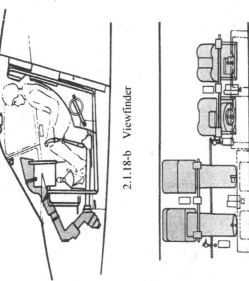

2.1.18-b Viewfinder

2.1.18-c Twin wide angle and narrow angle cameras

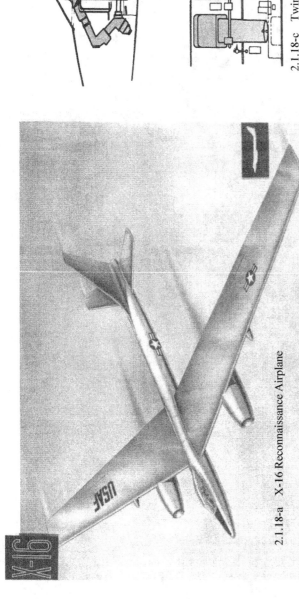

2.1.18-a X-16 Reconnaissance Airplane

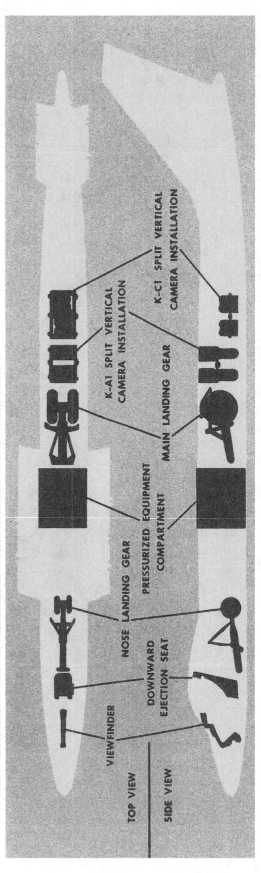

K-A1 SPLIT VERTICAL
CAMERA INSTALLATION

K-C1 SPLIT VERTICAL
CAMERA INSTALLATION

MAIN LANDING GEAR

PRESSURIZED EQUIPMENT
COMPARTMENT

NOSE LANDING GEAR

DOWNWARD
EJECTION SEAT

VIEWFINDER

TOP VIEW

SIDE VIEW

2.1.18-d Layout of equipment in X-16 airplane

2.1.18-e   Features of Bell Aerospace's X-16 reconnaissance airplane

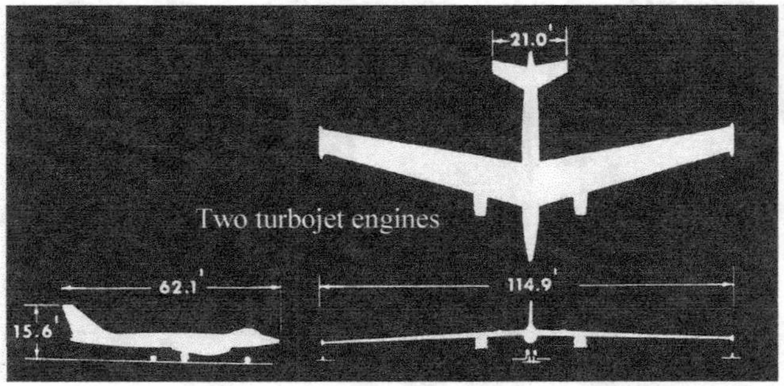

Two turbojet engines

Wing tip outrigger wheels support
the wings during ground operations

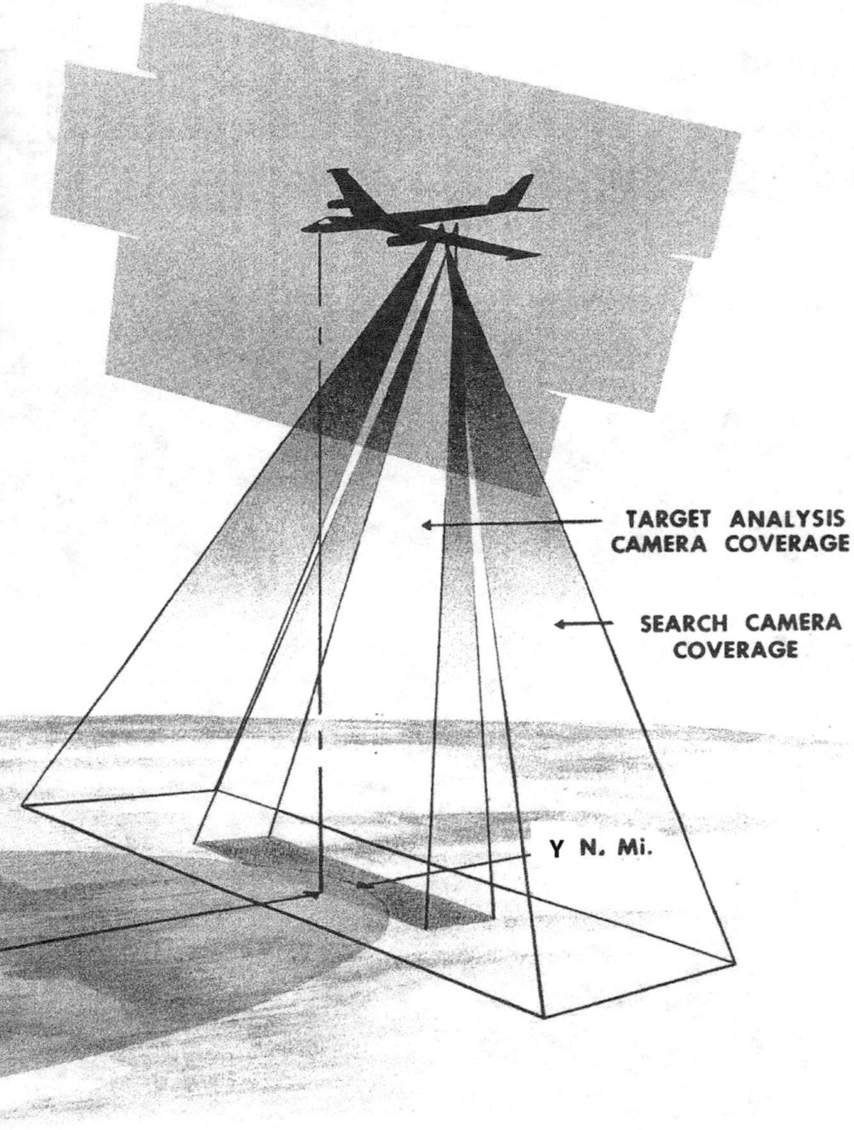

21.0'

114.9'

62.1

15.6'

TARGET   ANALYSIS
CAMERA   COVERAGE

SEARCH   CAMERA
COVERAGE

Y N. Mi.

X N. Mi.

VIEWFINDER COVERAGE

2.1.18-f Mockup of Bell Aerospace X-16 Reconnaissance Airplane

2.1.19　Bell's S-1 Supersonic Research Airplane/Missile

At that time, airplanes and missiles had not been flown extensively above 50,000 feet and 650 mph. The S-1 was developed to explore this high speed and high altitude region. The S-1 could take off and land from the ground or, to extend performance, it could be dropped, at high altitude, from a mother ship. A significant amount of S-1 space was reserved for specialized research equipment The craft could be operated as a piloted airplane or as an occupied missile. The S-1 evolved from the XS-1.

TRANSONIC AND SUPERSONIC RESEARCH

Armament Research
Supersonic Aerodynamics
  Control and Stability
    Control Forces
    Control Effectiveness
    Control Devices Maneuverability Considerations
  Performance
    Drag Effects Lift Devices
    Available Lift Coefficients Various Airfoil Shapes
  Internal Dynamics of Ducts and Diffusers
Power Plant Research
  Subsonic and Supersonic Ramjets Ducted
  Rocket Development Pulse Jet Testing
  Propeller Research at Supersonic Speed
Missile Development
  Full Scale Tests -- Guidance Devices
  Full Scale Tests – Target Seekers
  Testing of Computers
  Development of Tracking Methods
  Use as Maneuvering Targets for Testing of Other Missiles
  Use as Launcher for Very Small Guides Missiles
Aeromedical
  Personnel Escape at Supersonic Speeds
  Temperatures at High Speed
Missile Development
  Full Scale Tests-- Guidance Devices
  Full Scale Tests Target Seekers
  Testing of Computers
  Development of Tracking Methods
  Use as Maneuvering Targets for Testing of Other Missiles
  Use as Launcher for Very Small Guided Missiles

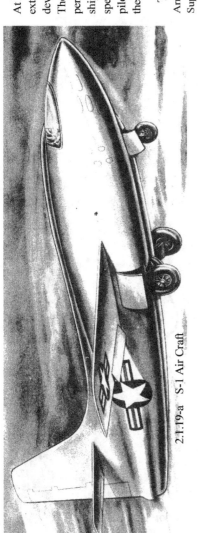

2.1.19-a　S-1 Air Craft

2.1.19-b　Flight Characteristics of S-1Supersonic Air Plane

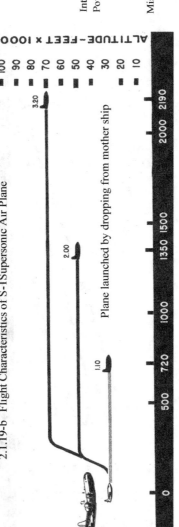

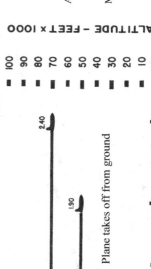

Plane launched by dropping from mother ship

Plane takes off from ground

PLANE SPEED IN MILES PER HOUR

84

FLIGHT TEST TIME

With 1,000 gallons of internal fuel, the S-1 can provide relatively extended test times, at altitude – an extremely attractive feature when compared with the limited transient data transmitted by missiles. The reusability of the S-1, as compared to that of an unmanned missile, makes it even more attractive for missile research.

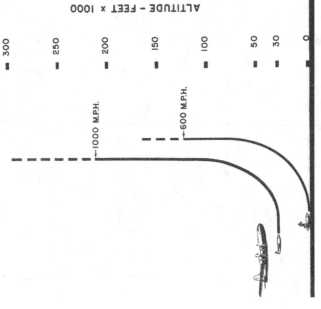

ALTITUDE - FEET x 1000

— 300

— 250

— 200   1000 M.P.H.

— 150

— 100   600 M.P.H.

— 50

— 30

0

2.1.19-d   Very high altitude research

VERY HIGH ALTITUDE RESEARCH

Cosmic Radiation Measurements
Physics of the Upper Atmosphere
Development of Aeromedical Devices
Communication in the Ionosphere
Proof Testing of Electrical Equipment
Photography and Mapping
Develop Methods of Measuring Very High Altitude Personnel Escape Devices

TIME - MINUTES   0   10   20   30   40   50

ALTITUDE-FEET x 1000   10   20   30   50   60   70   80

7.20 MIN.   10.70 MIN.   15.50 MIN.

Plane launched by dropping from mother ship

Pilot takes off from ground

4.10 MIN.   6.20 MIN.   8.10 MIN.

ALTITUDE-FEET x 1000   10   20   30   40   50   60   70   80   0   10   20   30   40   50

ALTITUDE-FEET x1000   35   70   105   140   143

27 MILES   M = 3.75

MILES   0   50   100   150   200   250   300   350
MINUTES   10   20   30   40   50

2.1.19-c   Flight test time

Remotely controlled airplane or missile strikes at a ground installation that is 35 miles from the air drop site. The plane successfully escapes, after dropping payload, and rendezvous with mother ship 55 miles away.

Air launched S-1, flying at Mach 1.2, attacks incoming enemy air craft with enough fuel for 1 minute of combat, at a distance of 95 miles from the air drop position.

S-1, taking off from a conventional runway, attacks enemy air craft at Mach 1.2, with enough fuel for 1 minute of combat, at a distance of 40 miles from the airstrip

Reference

(1) *S-1 Research Airplanes and Tactical Airplanes*, Bell Aerospace Corporation

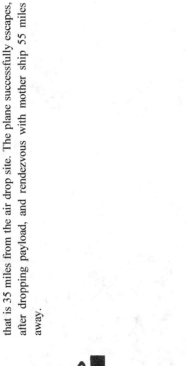

Distance (miles)

ALTITUDE - FEET × 1000

MAXIMUM DISTANCE FOR RETURN

SUFFICIENT FUEL FOR FIVE MINUTES CIRCLING AT ORIGINAL LAUNCHING POINT

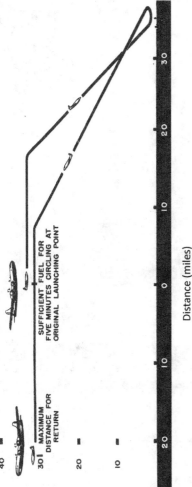

ALTITUDE

SUFFICIENT FUEL FOR 1 MINUTE OF COMBAT AT M=1.20

600 M.P.H.

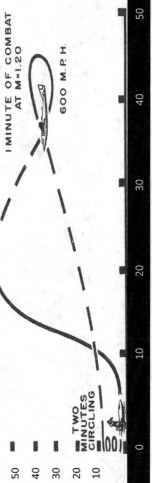

1 MINUTE OF COMBAT AT M=1.20

600 M.P.H.

TWO MINUTES CIRCLING

2.1.19-e   S-1 Combat missions

86

2.1.20   Assembled Bell Airplanes and Crews

2.2 Air Cushion Landing Gear for Airplanes

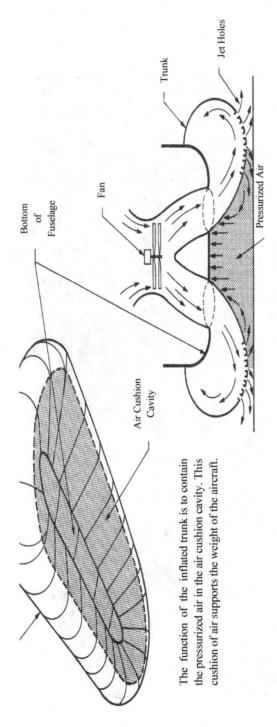

The function of the inflated trunk is to contain the pressurized air in the air cushion cavity. This cushion of air supports the weight of the aircraft.

2.2-a Air Cushion Landing Gear Principal

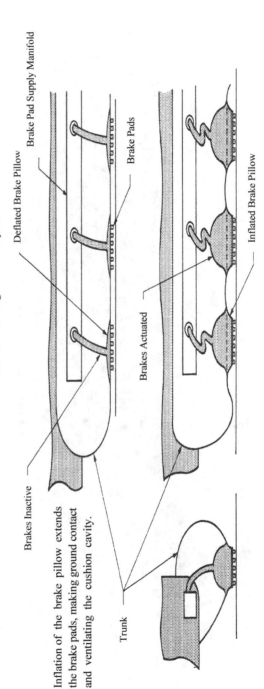

Inflation of the brake pillow extends the brake pads, making ground contact and ventilating the cushion cavity.

2.2-b Brake system for air cushion landing gear

2.2-c First air cushion landing of XC-8A -- April 11, 1975

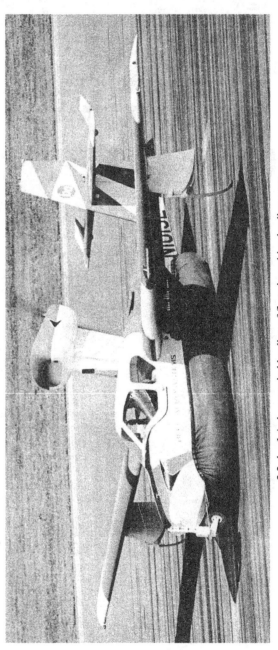

2.2-d LA-4 crosswind landing of first air cushion landing gear

89

2.2-e   Bell's Data Sheet for air cushion landing gear installed on LA-4 test airplane

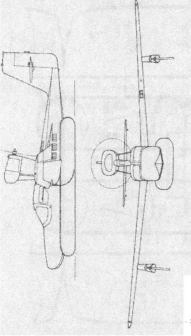

Air Cushion Landing Gear (ACLG) provides a superior replacement for conventional shock strut landing gear currently used on most aircraft. It replaces wheels, skis, or hull and floats on any size or type of aircraft with a single system combining the functional capabilities of them all.

The high degree of energy absorption, increased safety and reliability, weight savings and overall economy all contribute to ACLG system superiority.

This net application of the air cushion principle, as developed by Bell Aerospace, is an outgrowth of their work in the field of Air Cushion Vehicles (ACV). Incorporating the same amphibious characteristics as the ACV the aircraft using this system, with its new landing and takeoff techniques, has unprecedented capabilities,

The ACLG consists primarily of a doughnut shaped ban inflated to a thickness of approximately two feet, by axial fan. The fan, powered by a separate engine, forces air down into the bag. This flow of air escapes through thousands of tiny nozzles on the underside of the bag providing a cushion of air upon which the aircraft floats. The bag is elastic, fabricated from multiple layers of stretch nylon cloth for strength, natural rubber for elasticity, coated with neoprene for environmental stability. When deflated it fits snugly against the underside of the aircraft.

At touchdown, six brake skids on the underside of the ACLG bag are brought into contact with the landing surface by pneumatic pillows. When fully inflated for maximum braking, these pillows are each slightly larger than a basketball. For parking on land or water a lightweight internal bladder seals the air jets, thus, supporting the aircraft at rest, or providing buoyancy -- to keep it afloat indefinitely.

The ACLG minimizes airstrip requirements and enables aircraft to take off and land on unprepared surfaces: in open water, ice, snow, marsh, sand or dirt. Factors contributing to this feasibility: the very low pressure in the bag supporting the aircraft, the elimination of friction because of the air jets, and the increased area of contact during braking.

ACLG has proved to be an ideal gear for crosswind takeoff and landing. Liftoff and touchdown crabbed is no problem with the air cushion.

DIMENSIONS

Aircraft		
Wing Span		38 feet
Overall Length		24 feet 11 inches
Gross Wing Area		170 square feet
Air Cushion		
Length		16 feet
Width		3 feet 10 inches
Area		45 square feet

LOADINGS

Wing Loading		15 psf
Air Cushion Pressure		55 psf

WEIGHTS

Gross Operating Weight		2500 pounds
Air Cushion System Weight		258 pounds

POWER PLANTS

Propulsion Engine		Lycoming Model 0 360 01A
Rating		180 brake horsepower
Air Cushion Engine		Modified McCulloch Model 4318F (driving 2-stage axial fan)
Rating		90 brake horsepower

PERFORMANCE

Cruising Speed		125 mph
Stalling Speed		54 mph
Take-Off Run		650 feet
Landing Run		475 feet

2.2-f   Air cushion landing gear concept for military cargo plane.

During flight, the deflated air cushion bags are stored in the aircraft. Prior to landing, two axial fans draw in outside air to deploy and inflate the airbags, creating an air cushion 11.5 feet wide X 34 feet long.

After inflation, valves in the supply duct, are turned off to prevent backflow. Upon landing, the reinforced bag landing strips drag along the ground.

The bags are fabricated from a special Bell developed material, 0.25 inch thick; three layers of nylon cloth are situated between four layers of rubber. Thousands of small vent holes crisscross the bottom of the air bags. The material allows the bags to stretch to triple it width; it cannot be stretched lengthwise.

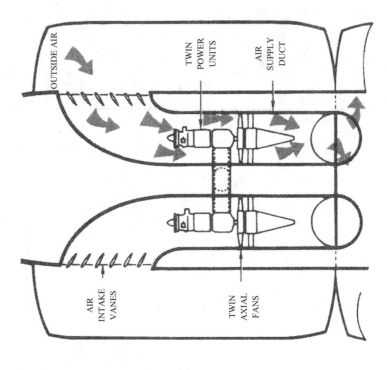

Top view showing outside air being drawn into air cushion bags

Air cushion bags deployed during landing of military cargo plane

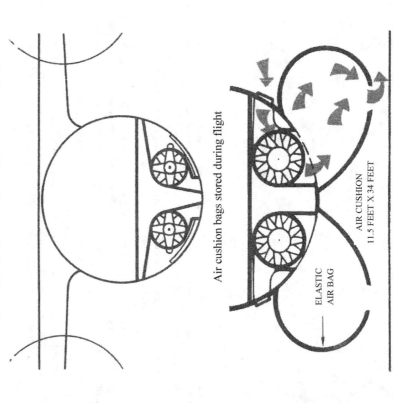

Air cushion bags deployed during landing

2.3 Vertical/Short Takeoff and Landing (V/STOL) Airplanes

2.3.1 Bell's V/STOL "Air Test Vehicle"

2.3.1-a Bell's Data Sheet for V/STOL "Air Test Vehicle"

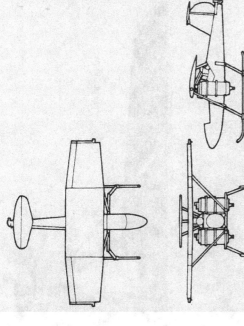

DIMENSIONS:
Length   20.7 feet
Height   7.7 feet
Span   26.0 feet
POWER PLANTS:
Two Fairchild J-44-R1 Turbojets
(1,110 pound thrust each)
One Palouste turbo-compressor
WEIGHT:
Gross   2046 pounds
PERFORMANCE:
Maximum Speed   100 mph

World's first jet-powered V/STOL was the Bell Air Test Vehicle(ATV) which flew in 1953.

This pioneering concept was powered by two J44 turbojet engines, attached to either side of the fuselage, which rotated from vertical to horizontal in transitioning from liftoff to conventional flight. Its fuselage was a Schweizer glider with a wing and tail from a Cessna. A separate Palouste turbo compressor supplied air for jet reaction controls. The ATV was purely a company sponsored research tool which demonstrated the adequacy of a compressed air reaction control system to provide acceptable VTOL pilot handling qualities. It served to establish the basic feasibility of jet VTOL.

Fabrication of the Air Test Vehicle was completed in seven months. The ensuing flight program demonstrated conclusively the technical practicality of the rotating thrust line, horizontal attitude V/STOL concept.

Today it is enshrined along with the many other "firsts" of American aviation in the Smithsonian Institution, Washington, D.C.

BELL AEROSYSTEMS
BUFFALO, NEW YORK   A Textron COMPANY

2.3.1-b   Close-ups of Bell's "Air Test Vehicle" (ATV) – the world's first jet powered V/STOL airplane

Side/front view

Pilot controls rotation of turbojet engines from vertical (lift-off) to horizontal (level flight)

Front view

93

2.3.1-c Bell's "Air Test Vehicle" taking off vertically and then changing to horizontal flight

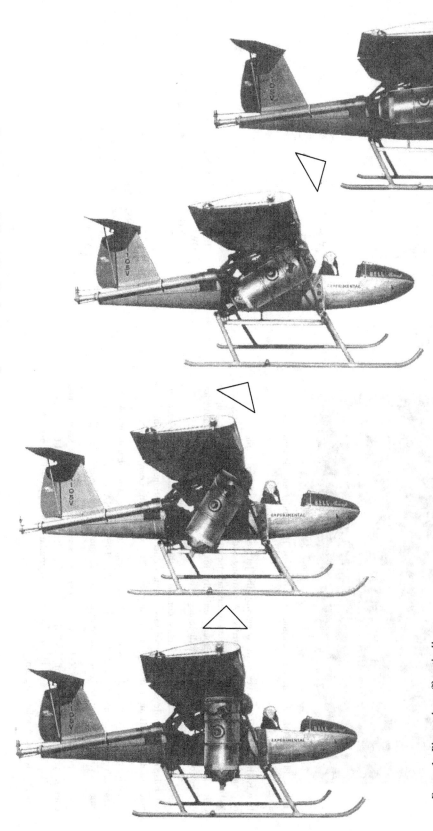

Frame 1: Plane takes off vertically.

Frames 2 & 3: Engine starts rotating. Plane continues to climb but also begins horizontal movement.

Frame 3: When engines are rotated horizontally, plane flies at constant altitude.

2.3.2 Bell's X-14 V/STOL Vectored Jet Airplane

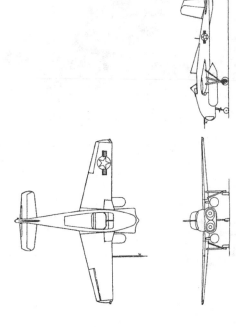

2.3.2-a   Bell's Data Sheet for X-14 V/STOL airplane

DIMENSIONS:
Length   25 feet
Wing Span   34 feet
Tail Height   8 feet
POWER PLANTS:
Two General Electric J85-5 turbojet engines
(2,859 pounds thrust each)
WEIGHT:
Gross   4,000 pounds
PERFORMANCE:
Operational Speed   160 knots
Maximum Speed   180 knots
CAPACITY:
1 pilot

The X-14 V/STOL aircraft was designed and built by Bell under an Air Force contract awarded in July 1955.

It was the first VTOL ever built to employ the jet vectored thrust principle. Powered originally by twin, nose-mounted British Armstrong-Siddeley Viper turbojet engines, the X-14 achieves flight by means of rotating cascade thrust diverters mounted at the tailpipe exit. The pilot directs thrust either vertically for hovering flight, horizontally for conventional flight, or at an intermediate angle for transition.

During hovering and low speed flights, control of the aircraft is maintained by reaction controls. Aerodynamic surfaces serve this function during conventional flight. This two-place, mid-wing, all-metal monoplane is equipped with fixed landing gear.

The aircraft was delivered to National Aeronautics and Space Administration's Ames Research Center, Moffett Field, Calif., in October 1959. NASA replaced the original engines with J85 turbojets for increased thrust and designated the vehicle the X-14A.

Primary purpose of NASA's X-14A program is to research and define the stability and control system requirements of V/STOL aircraft. In addition it has been used for V/STOL test pilot familiarization and to investigate and simulate the approach phase of lunar landings for the Apollo program.

BELL AEROSYSTEMS
BUFFALO, NEW YORK   A textron COMPANY

95

2.3.3 Bell's D-188A V/STOL Fighter/Bomber Jet Airplane

2.3.3-a Bell's Data Sheet for D-188A V/STOL Fighter/Bomber airplane

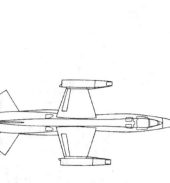

The D-188A or XF-109 was designed by Bell as a fighter-bomber weapon system under a joint Navy/Air Force contract and developed between 1957 and 1959.

Although it progressed only through the mockup stage, the concept has been proven successful since in the VJ-101C aircraft built by a West German development group and currently flying in Europe.

During the U.S. program, a complete set of design analysis layout drawings and subsystem specifications were prepared and extensive wind tunnel, jet impingement, aeroelastic, and structural testing programs were completed. The contractual efforts were completed and the mock-up reviewed by Navy and Air Force personnel in February 1959.

The single-place D-188A resembled a conventional jet fighter. It had a long wasp-waisted fuselage and short knife-like wings. Powered by eight J85 turbojet engines, two mounted at the tip of each wing, four located in the fuselage, it was designed as a deck-ready interceptor for the Navy and as a tactical fighter-bomber for the Air Force.

Wing-tip engines rotated to a vertical position for takeoff, supplemented by two lift engines in the forward fuselage and by diverting the thrust from the rear fuselage engines, Transition to horizontal flight could be made in 60 seconds.

DIMENSIONS:
Length   58.6 feet
Width   23.8 feet
Height   12.9 feet
POWER PLANTS:
      Eight J85 turbojet engines
WEIGHT:
      Gross   31,000 pounds
PERFORMANCE:
      Maximum Speed   Mach 2.3
CAPACITY:
      1 pilot
      2,000 pounds payload

BELL AEROSYSTEMS

BUFFALO, NEW YORK

A textron COMPANY

96

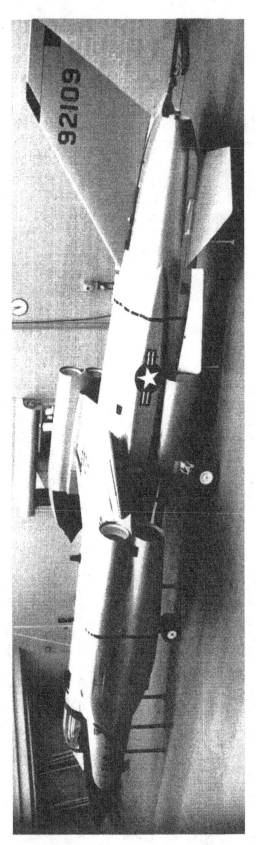

2.3.3-b   Close-up of D-188A V/STOL airplane

2-3.3-c   Phantom view of D-188A V/STOL airplane

Two rotatable J85 turbojet engines
attached to each wing tip

97

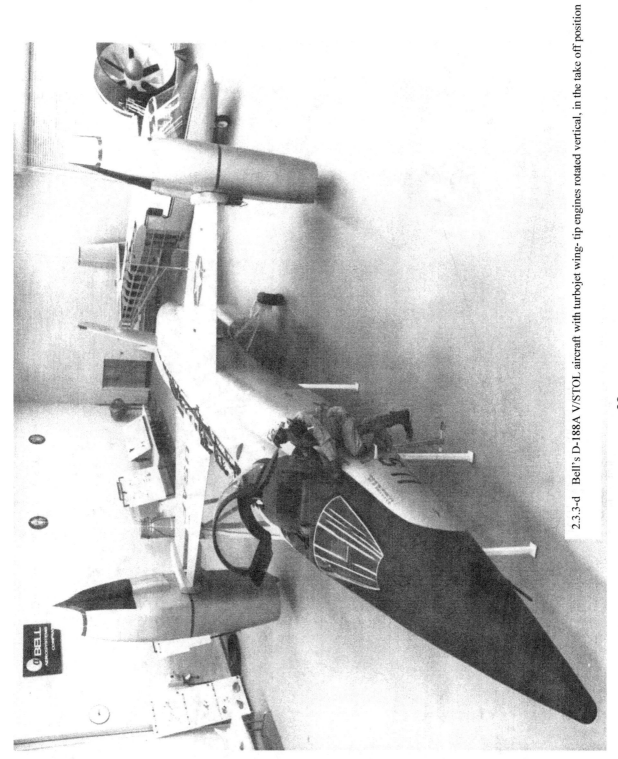

2.3.3-d   Bell's D-188A V/STOL aircraft with turbojet wing- tip engines rotated vertical, in the take off position

98

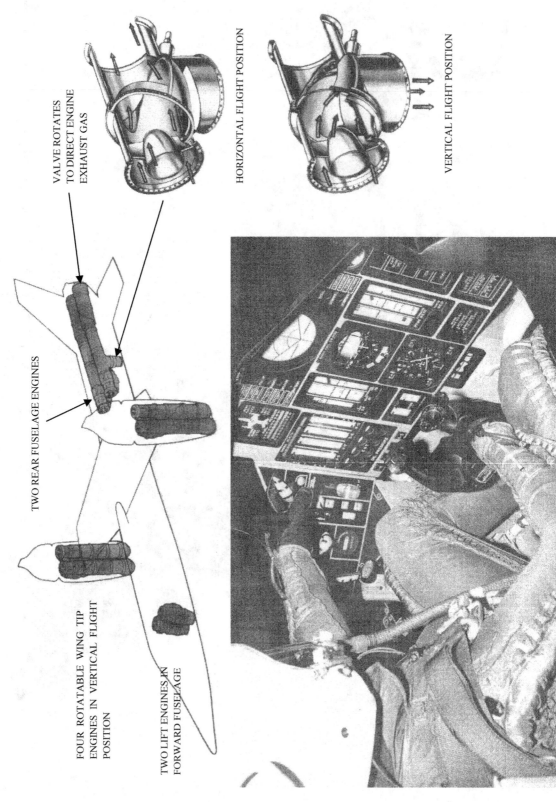

VALVE ROTATES
TO DIRECT ENGINE
EXHAUST GAS

HORIZONTAL FLIGHT POSITION

VERTICAL FLIGHT POSITION

2.3.3-e   Location of eight J85 turbojet engines in D-188A V/STOL airplane

TWO REAR FUSELAGE ENGINES

FOUR ROTATABLE WING TIP
ENGINES IN VERTICAL FLIGHT
POSITION

TWO LIFT ENGINES IN
FORWARD FUSELAGE

2.3.3-f   Pilot in cockpit of D-188A V/STOL airplane

99

2.3.4 Bell's XV-15 V/STOL Airplane

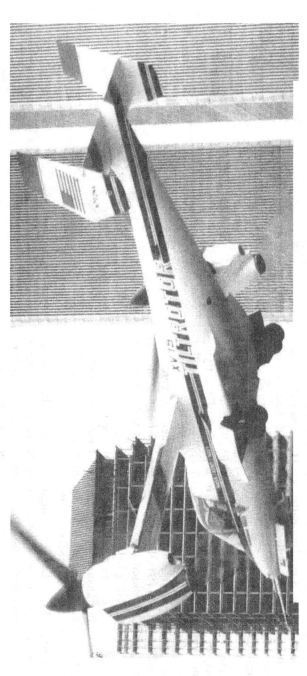

2.3.4-a Bell XV-15 V/STOL airplane has the propeller mounted directly to the engine, at the tip of each wing. To transition between vertical and horizontal flight, the entire motor/propeller assembly is rotated. First flight May 3, 1977.

2.3.5 Bell's XV-3 V/STOL Airplane

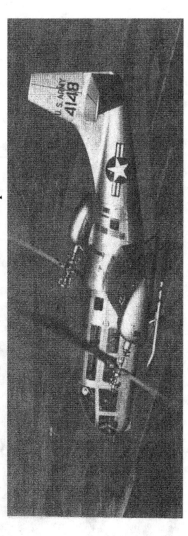

2.3.5-a Bell XV-3 V/STOL airplane. Engines are mounted in the fuselage and drive shafts transmit power to the propellers, which can revolve to provide vertical take-off and landing, -- like a helicopter. 1959.

2.3.6 Bell's X-22A V/STOL Research Airplane

2.3.6-a Bell's Data Sheet for X-22A V/STOL research airplane. On May 25, 1964 the last military aircraft was built in Western New York – an X-22A V/STOL craft.

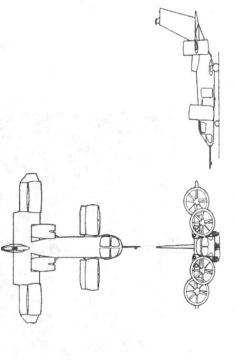

The X-22A is a dual tandem, ducted propeller V/STOL research aircraft designed and built by Textron's Bell Aerospace Company under a Navy administered contract.

An important feature of the X-22A is its extremely versatile variable stability and control system (VSS). The VSS enables the X-22A to conduct flight research on the handling qualities for this type of V/STOL and for . simulating a variety of V/STOL designs, permit investigation of the flying characteristics and flight control problems generally applicable to all other V/STOL aircraft.

The X-22A made its maiden flight on March 17, 1966, when it hovered for 10 minutes. A short takeoff and (STOL) was first accomplished on June 30. The first vertical takeoff, transition to conventional flight and return to a vertical landing, took place March 1, 1967.

For takeoff, the X-22A's ducts are rotated to a vertical thrust position. As altitude is gained, they are transitioned to a horizontal thrust position for forward flight. For landing, the procedure is reversed. For STOL takeoff or landing, the ducts are set at an intermediate position.

Four General Electric T58 turbo shaft engines, rated at 1250 shaft horsepower, provide the power to drive the seven foot Hamilton Standard propellers. Power is transmitted from the engines to the propellers through a system of shafts and gearboxes, so interconnected that a single engine can turn all four propellers.

A test program was underway to demonstrate this dual tandem, ducted propeller aircraft and to evaluate the military potential of this new concept of flight.

DIMENSIONS:

Length	39.6 feet
Height	20.7 feet
Span	39.2 feet

POWER PLANT:
Four YT58-GE-8D turboshaft engines (1250 hp each)

WEIGHT:

VTOL Gross	16,274 pounds
	(Standard day – one engine out)
Max Gross	18,016 pounds
	(STO over 50-720 feet)

PERFORMANCE:

Speed		322 mph
Endurance		
	VTOL	2.9 hours
	STOL	4.4 hours
Range	VTOL	455 n.m.
	STOL	685 n.m.
Hover ceiling		11,000 feet
	(Standard day – 4 engines)	

PAYLOAD:
Two crew; 1500 pounds research instrumentation (including VSS), cargo, or six passengers.

Bell Aerospace
DIVISION OF textron

BUFFALO. NEW YORK 14240

101

2.3.6-b Propulsion test stand for X-22A V/STOL airplane

Bell began working with, in 1953, ducted propellers that rotated to give a V/STOL capability. Capturing the interest of the U.S. Army, they received a contract for a design study to determine the feasibility of the ducted propeller for V/STOL airplanes. The concept was found to be viable. Since then, Bell has extended its work to: the building of a full scale mockup, the testing of propeller models, model testing in a wind tunnel, and simulations employing analog computers.

This work has been extend to the construction of a test stand for evaluating a full scale model of four ducted seven foot propellers powered by four General Electric T58 turbo shaft engines. The stand utilized many of the parts that will be used in the final airplane – the engines, steel and fiberglass propellers, gear boxes and shafts. The engines are capable of driving the propellers between 1,800 and 2,600 rpm, with the propellers on

the right spinning clockwise and those on the left spinning counter-clockwise. The propeller pitch angle can be changed, as can the orientation of the ducts/ propellers.

Gear boxes and shafts are installed to interconnect all the engines and propellers. In the event of one or two engine failures, the craft will remain airborne since the remaining engines will drive all four propellers.

Multiple test objectives were identified: (1) demonstrate that the design is acceptable, (2) develop confidence in the transmission system, (3) conduct an extensive vibration analysis, (4) identify acceptable service and maintenance procedures.

In forward flight, the duct acts like a circular wing. It increases lift beyond what the wings can create. The ducts also help to reduce propeller noise – the main noise source – in two ways. For a given required thrust the propellers can rotate slower, which reduces noise. Also the duct itself absorbs some of the noise.

The ducts increase ground safety. They prevent personnel from inadvertently walking in to the propellers and prevent objects from falling into, or being thrown into, the propellers.

The next step was to build a model for wind tunnel testing. These test results guided the development of the first full size ducted plane – the X-22A. With the X-22A, it was shown that the ducts actually increased the vertical take-off thrust from 3,770 pounds to 5,000 pounds, per duct.

Design of the X-22A allows two ways of controlling the plane in, both forward or hover flight:

(1) Elevons, mounted on the outlet side of the ducts, can be rotated to change the direction of the exhaust jet and resulting thrust

(2) The propeller pitch can be changed on each propeller to alter the thrust from each engine

Propeller pitch and elevon rotation are both controlled by the pilot, in pairs – either left and right pairs of forward and aft pairs.

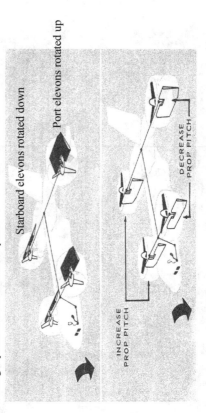

Starboard elevons rotated down

Port elevons rotated up

INCREASE PROP PITCH

DECREASE PROP PITCH

2.3.6-f   X-22A roll control using elevons or propeller pitch

Reference

(1) Bissel, J., Henderson, C., *Why the X-22A's Propellers Have Ducts*, Rendezvous, Bell Aerosystems, Vol. VI, No. 5, 1967.

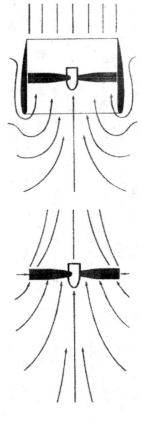

2.3.6-c Airflow through rotating propeller when duct has been removed   2.3.6-d Airflow through rotating propeller when installed in a duct

While the installation of the propellers inside a cylindrical duct is a bit unconventional, there are reports that this configuration was actually seriously investigated in 1927. Bell, however, didn't start considering this approach until 1953. Thereafter, the Navy became interested in possible application to assault transports. Under contract, in 1957, Bell designed, built, and help test a full scale ducted propeller. Extensive tests identified the effects of the ducts on thrust, drag, lift, and pitching moments. It was determined that the ducted propeller creates more thrust than an un-ducted propeller, of equal horsepower – when the plane is taking off vertically or moving slowly.

According to Figure 2.3.6-c and 2.3.6-d, this partially occurs since significantly more air flows uniformly through the propeller, when the duct is installed, because the duct prevents air flow loss out the end of the blades.

Figure 2.3.6-e illustrates another reason, when the propeller blades are oriented horizontally. Near the duct wall, air is drawn up vertically along the outside surface of the duct. As the air passes around the end of the duct, and into the duct, it speeds up. Using Bernoulli's fluid flow equation, it can be shown that the air pressure will drop as the flow velocity increases. With normal atmospheric pressure on the rest of the duct, there is a net upward force on the duct, which contributes to the observed increased lift. Half the lift comes from the propeller and half from the duct producing 26% more lift than with no duct.

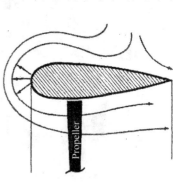

Propeller

2.3.6-e Airflow around end of duct with duct oriented vertically

2.4.1 History of the Development of Helicopter Technology by the Bell Aircraft Company

1928: Arthur M. Young began looking for some type of a challenging project. He eventually selected development of a "rotary wing machine". To date, no "rotary wing machine" had ever been successfully flown. For the next thirteen years he experimented with various types of models of "rotary wing machines"; see Figure 2.4.2-a.

Sept. 3, 1941: Young demonstrated his flying model to Larry Bell and his engineers.

Nov. 24, 1941: Arthur Young began working for Bell Aircraft Company to supervise the design and construction of several full size prototype helicopters.

June 1942 to June 1945: First helicopter plant opened in Gardenville, N.Y. to build prototype helicopters.

Unknown Date: Even though he wasn't a pilot, Arthur Young successfully "flew" the first helicopter, a tethered Bell Model 30-1, to an elevation of one foot.

June 26, 1943: Experienced pilot Floyd Carlson, after teaching himself how to fly a helicopter, made the first un-tethered flight of Bell Model 30-1A, around a field behind the factory.

May 10, 1944: Bell Model 30-2 put on a public demonstration inside the armory in Buffalo, N.Y.; Figure 2.4.3-d.

1945: The Bell Model 47 entered military service with a designation of **H-13 Sioux**, where it could support a crew of three. It could be armed with twin M37C 30 caliber machine guns or twin M-60 machine guns. It was used for medical evacuation with two liters, reconnaissance, and observation.

Dec 1946: Bell Model 42 was displayed at the Cleveland Air Show. It was large enough to carry two crew members up front and three passengers in the back. The frame had an all metal skin. The cabin was entered through two car like doors. Three prototypes were built.

March 8, 1947: Bell Model 47 received the first Civil Aeronautics Administration certification for a helicopter. The Model 47 would be manufactured from 1946 to 1973, in 20 different variants, with a combined civilian and military production of 5600; Figure 2.4.3-j.

1948: Bell Model 54, designated by the U.S.A.F. as **XH-15** was developed as a four seated all metal helicopter with tricycle landing gear and a single two bladed main propeller. Three craft were built.

May 13, 1949: A Bell Model 47 set an altitude record of 18,550 feet

September 21, 1950: A Bell Model 47 became the first helicopter to fly over the Alps.

1951: Relocation of Bell Helicopter Division to Texas.

September 17, 1952: By flying nonstop from Hurst, Texas to Buffalo, New York a Bell Model 47 set a world distance record of 1,217 miles.

March 4, 1953: Bell Model 61, designated as **HSL-1**, developed as an antisubmarine helicopter to search for and destroy submerged enemy submarines. Proved to be so noisy that sonar operators had a hard time hearing return signals. It also proved to be too big to fit onto aircraft carrier elevators, even with the blades folded. The only twin blade helicopter developed by Bell; Figure 2.4.4-a.

1954: A variant of Bell Model 47, designated as the **XH-33** and **XV-3**, was developed as a V/STOL craft.

1955: Bell Model 201, designated as **XH-13F**. A model 47G was fitted with a turbine and used to evaluate turbine performance, because of its advantages.

1956: Bell Models 204/205. Civilian versions of the military UH-1 Iroquois single engine helicopter. Used for fire fighting, cargo lifting, and crop dusting.

October 1956: Military version of Bell Model 204, officially named the Iroquois, was designated as a utility helicopter, the **UH-1**, the first turbine powered helicopter. It was commonly called the **"Huey"** after its UH-1 designation. Used extensively for front line casualty evacuation in Korea.

1962: Bell Model 207 – the **Sioux Scout**, a variant of Model 47, was developed as a gunship.

August 10, 1962: Bell Model 533 – A research helicopter, a modified version of the UH-1 Iroquois, built for the U.S. Army to study helicopter improvements. Given the designation YH-40. Obtained a maximum speed of 316 mph.

Sept. 10, 1971: Bell Model 309 **King Cobra**. A company funded program that developed two prototype craft as a proposal for the next generation attack helicopter that improved high speed performance and reduced noise. Introduced numerous advanced features: night vision TV, inertial navigation system, infrared fire control system, day and night laser sighting, autotrim system, and an automatic stabilization system.

1972: Bell Model 214 medium lift helicopter. Uses a single engine that's more powerful than the Model 205. To improve lifting capacity and performance, at high temperature and high altitudes, an upgraded rotor system was adopted.

October 1, 1975: Bell Model 409, designated by the US Army as **YAH-63**. Two craft were built as experimental attack helicopters. It utilized: two bladed propeller, tricycle landing gear, antitank missiles mounted on short wings attached to each side of the lower fuselage, and a 30mm cannon mounted in a rotatable chin turret that was installed just below the nose. The two man crew was seated with the pilot in front of and below the co-pilot, in order to improve pilot visibility.

1977: Bell Model 222, a midsize twin turbine helicopter. A number of sophisticated features were introduced: the retractable landing gear was stored in new stub wings, a Noda Matic system was installed to reduce vibration, and safety was improved by installation of dual hydraulic and electrical systems.

1978: Bell Model 214ST, a stretched version of the Model 214 that seats 19.

1979: Bell Model 412, a utility helicopter derived from the Model 212. It's main difference was that it employed a four bladed propeller.

January 18, 1981: A Model 222 was delivered, the 25,000th helicopter manufactured by Bell.

1984: Bell Model 406CS **Combat Scout**. Used for: painting targets with a laser for Apache Hellfire helicopter, artillery spotting, and aerial reconnaissance. Various sights could be mounted on the roof.

1985: Bell Model 412SP special performance version of the Model 412. Upgrades included a larger fuel tank, higher take-off weight, and more seating capability.

*Engineering at Bell Helicopter Company, Bell Aircraft Company, Unknown Date.
*Voss, Carrol and Moore, C.W., The Bell 47B-3 Helicopter, Bell Aircraft Co., 1947

Cenkner Table

105

1963: Bell Model 206 developed as a four seat light observation helicopter for the U.S. Army Light Observation Program. Only four were built.

June 27, 1963: Bell Model 207 Sioux Scout A development helicopter that had the goal of improving the performance of armed combat helicopters. Introduced various features that are routinely used today: chin turret with two 7.62 mm machine guns, stub wings to give extra lift and increase range by carrying extra fuel, dual flying controls, integrated weapons and sighting systems, low drag profile, and crew of two seated in tandem with the pilot above and behind the co-pilot/weapons officer.

March 27, 1965: Bell Model 208 developed by fitting two turbines on a UH-1D helicopter.

1965: Bell Model 209, developed under a company sponsored program, was the jumping off point for the development of the Bell **AH-1 Huey Cobra**. The AH-1 was developed as the first armed battlefield helicopter. It had a streamlined body with the pilot sitting above and behind the co-pilot/gunner. Its armament could include guided missiles, aerial rockets, air-to-air missiles, and a 20 mm cannon.

January 10, 1966: Bell Model 206A was a modification of the Model 206 airframe that became a five place commercial helicopter designated as the **Jet Ranger**.

1967: Bell Model 205D had its fuselage enlarged by 41 inches; was renamed the **UH-1D**. Total seating capacity, with crew, was increased to 15 while stretcher carrying capacity was increased to six.

1968: Bell Model 206A modified as U.S.N **TH-57 Sea Ranger** and U.S.A. **OH-58A Kiowa**. The Kiowa was an armed reconnaissance helicopter that was used to support troops on the ground. The Sea Ranger was primarily used to train Navy pilots.

1970: Bell Model 212, a twin turbine Huey, designated as the **UH-1N** by the military and **Model 212 twin two-twelve** for the commercial version, which could carry a pilot and fourteen passengers.

1971: Bell Model 240 -- a proposal for the US Army Utility Tactical Transport Aircraft System that was to replace the Bell Huey. The helicopter employed twin engines, four blade main propeller, and it could seat fourteen people.

2.4.2 Early Development of Helicopter Technology, by Arthur M. Young, Employing Small Scale Models

2.4.2-a   Arthur Young's small scale helicopter models

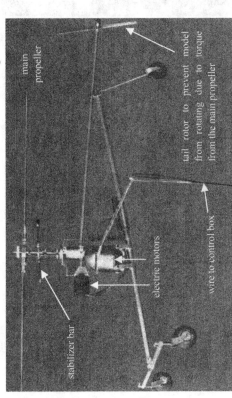

Arthur M. Young received a degree in mathematics from Princeton University in 1927. The next year he began searching for a challenging project. After some research, his attention was drawn to "rotary wing machines", partially because no one had been able to develop such a machine even though many had tried.

He established a laboratory, in a barn on his family's estate, and began experimenting with small models. His first models were powered with rubber bands but he eventually switched to electric motors. For the next thirteen years he experimented with small models that were more and more sophisticated.

His biggest problem turned out to be an inherent instability of the craft, which always culminated in its crashing. To stabilize the craft, he invented and patented the stabilizer bar – a long thin bar with a streamlined weight at each end. The bar was rigidly attached directly to the rotor, and flexibly attached to the mast, so that the bar could maintain the rotor plane horizontally regardless of the orientation of the mast; gyroscopic inertia was used to maintain the rotor horizontally.

His more sophisticated models used electric motors that could be remotely controlled, so that these models could be remotely flown to various heights (above figures).

On September 3, 1941 Arthur Young demonstrated his flying model to Larry Bell and a Group of his engineers. Before Bell would provide funding, he insisted on a demonstration that the helicopter would auto-gyrate, that is the propeller would continue to rotate and provide lift – so it could land safely – after the engine failed.

Young demonstrated the capability with his model, so Bell hired Young and his assistant to develop several full scale prototypes, which were designated Bell Model 30-1 and Bell Model 30-2. When the first model was severely damaged during a test, it was rebuilt and designated Model 30-1A.

On November 24, 1941 Arthur Young and his assistant Bart Kelly started working for Bell Aircraft Company to supervise the construction and test of the prototype helicopters at a new plant in Gardenville New York. At that time there were about fifteen employees at this plant.

References

(1) Tipton, Richard S., *They Filled The Skies: Arthur Young ... Maker of the Bell Helicopter and Larry Bell ... Aviation Trailblazer*, Bell Aircraft Company, Unknown date.
(2) AN *AVIATION STORY: THE HISTORY OF BELL AIRCRAFT CORP. 1935-1945*, Bell Aircraft Company, Unknown date.

2.4.2-c Arthur Young, in 1941, remotely controlling his helicopter model. His original laboratory was a barn.

Powered propeller at the tip of each rotating wing.

Four rotating wings.

2.4.2-b An early concept tested and discarded by Arthur Young. Each rotating wing had a small powered propeller, mounted at the tip, to rotate the wing.

2.4.3   Bell's First Helicopters – the Model 30 and Model 47

2.4.3-a   Controlled tethered flight of the first full
scale Bell helicopter, Model 30-1.

The pilot, Floyd Carlson, had to teach himself how
to fly a helicopter. Having started to fly at the age
of 14, he had a considerable amount of experience
flying fixed wing airplanes.

The tether was removed on June 26, 1943 so
Carlson could take it on its maiden flight around
the open field behind the helicopter factory.

Vibration problems were encountered when the
craft reached 25 miles per hour. Carlson suggested
putting a stiffening brace on the rotor, which solved
the problem.

The propeller could auto-rotate. If the engine
failed, the propellers would still rotate and provide
lift, so the craft could land safely.

Rotating wings provide lift

Stabilizer Bar

Tail rotor prevents
craft from rotating

Four landing skids

Tail rest

Ship 1, Model 30-1

2.4.3-b The helicopter factory in Gardenville New York, near
Buffalo and about 10 miles from Bell.

A Chrysler Agency was converted into a machine shop, office,
drafting room, and a workshop for manufacturing helicopter
blades. An adjoining garage served as the production facility.

The location was ideal because it had a large open field behind it,
where short flights could be made.

108

2.4.3-c Bell Aircraft's modified first helicopter, the Model 30-1A, had a 32-foot rotor and was powered by a 160 hp Franklin air-cooled engine. Wheels replaced the skids on Model 30-1, the frame was covered, and it was renamed Model 30-1A.

2.4.3-d The second prototype Bell helicopter, Model 30-2, had a radical design change; there was an intent to make it look more like a fixed wing airplane. On May 10, 1944 there was a public demonstration inside the armory in Buffalo, New York.

109

2.4.3-e   Arthur M. Young, the developer of Bell's helicopter technology, and his first
          full size helicopter, Model 30-1 (1943)

2.4.3-f   Engine and power train for the Model 47G-5 helicopter,
          typical for all of Bell helicopters

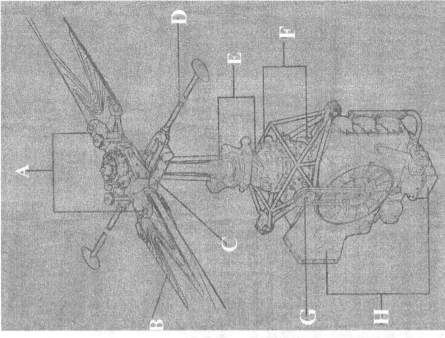

Stabilizer Bar

A. ROTOR HUB   E. SWASHPLATE ASSEMBLY
B. MAIN ROTOR BLADES   F. TRANSMISSION ASSEMBLY
C. STABILIZER ASSEMBLY   G. FAN DRIVE
D. STABILIZER BAR   H. ENGINE

A major problem that he encountered while testing his models of a "rotary wing machine" was
that the model became unstable and eventually crashed. To overcome this problem he invented,
and patented, the stabilizer bar. To allow control of the rotor plane, independent of the mast, the
bar was attached directly to the rotor.

110

2.4.3-g   Model 30-1A transmission and mast assembly being installed by
Arthur Young (right) and assistant Bart Kelly – Oct 1942

Model 30 – the thirtieth aircraft developed or proposed by Bell Aircraft Company, but
the first helicopter – had three different craft built during its development.

Ship No. 1
Built of 3 inch aluminum tubing   Used landing skids
160 hp Franklin air-cooled engine   Open cockpit
No covering on frame

Ship No. 1A
After ship no.1 crashed, it was rebuilt, modified, and designated ship no.1A.
The landing skids were replaced by four landing wheels
A blue skin was installed on the fuselage and the tail boom

Ship No. 2
Enclosed cabin with room for a passenger   Four landing wheels
Demonstrated to public in Buffalo Armory May 10, 1944

Ship No. 3
Different body shape   Bubble canopy
Instrument panel in middle, between pilot and passenger
Unobstructed view up and down because there was almost no floor

111

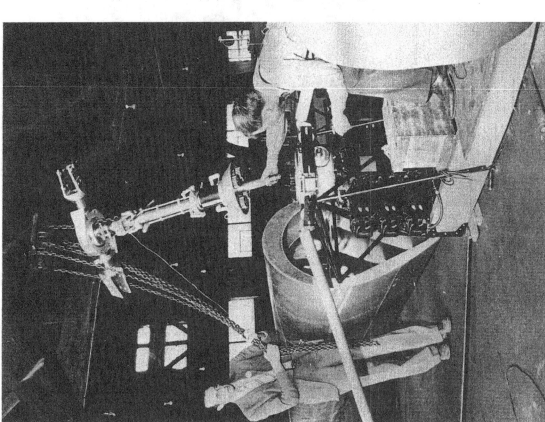

2.4.3-h   Bell's second helicopter, the Model 47, was outfitted for agricultural work. In 1947, it was successfully used to control a locust plague, in Argentina, by spraying pesticides on airborne and landed locust – a first.

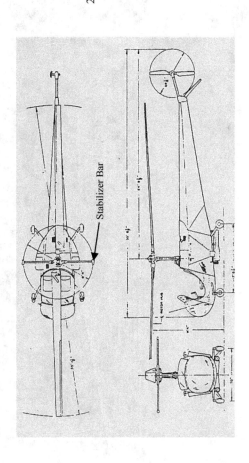

2.4.3-i   Bell Model 47 helicopter was developed from Model 30-3. Over the years, Model 47 had numerous variants. Model 47B-3, shown here, used an open cockpit, four landing wheels, and metal skin on the frame.

Stabilizer Bar

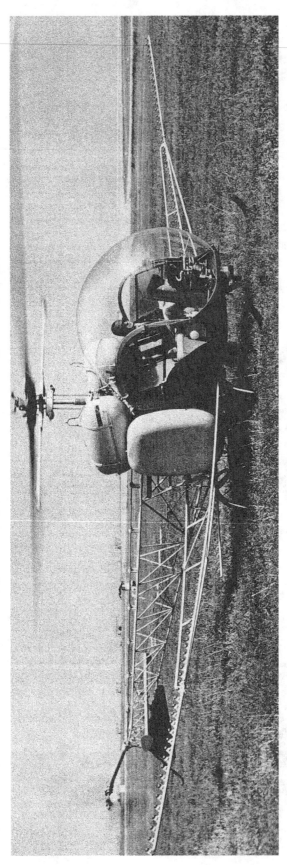

2.4.3-j   Bell Helicopter Model 47 G-5 modified for agricultural spraying

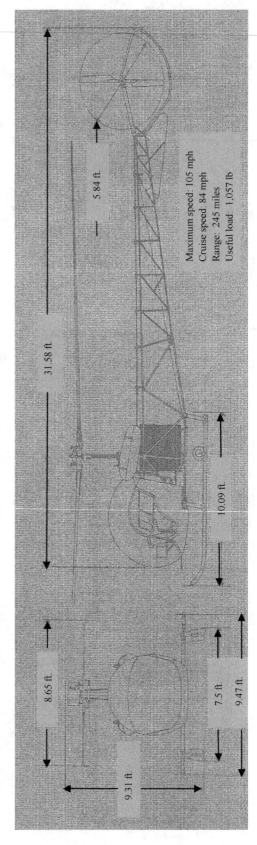

2.4.3-k   Specifications for Bell Helicopter Model 47 G-5

31.58 ft.

5.84 ft.

Maximum speed: 105 mph
Cruise speed: 84 mph
Range: 245 miles
Useful load: 1,057 lb

10.09 ft.

8.65 ft.

7.5 ft.

9.47 ft.

9.31 ft.

113

2.4.4 Bell's Only Twin Bladed Helicopter, the HSL-1

2.4.4-a   Bell's Data Sheet for the HSL-1 U.S. Navy helicopter, developed for submarine detection and destruction

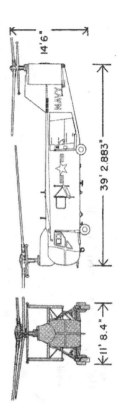

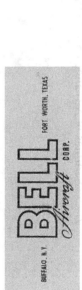

This is the first helicopter specifically designed for submarine detection. Developed for the U.S. Navy, the tandem-rotor craft marks Bell's first departure from its familiar single rotor helicopter configurations, incorporated on all Bell models from its first experimental helicopter in 1942. Promising to be the most formidable tactical aircraft or its type, the HSL-1 incorporates all the essential equipment necessary for submarine detection and destruction. It features compactness in size in combination with high rotor disc loadings for greater speed and range. The fore and aft rotors are inter-connected and power is supplied by a Pratt & Whitney R-2800 engine. To facilitate storage aboard an aircraft carrier or other type of ship, the rigid two-bladed rotors can be folded to reduce the overall length. The HSL-1 is truly capable of all-weather operation because of its high degree of stability in combination with a practical autopilot, also developed by Bell.

114

2.4.5 Samples of Advanced Single Propeller Helicopters Developed By Bell

2.4.5-a   Bell Model 214ST, Royal Thai Navy

ST – Super Transport
Large five blade rotor – 15.85 m diameter
2x General Electric CT7-2 turboshaft engine rated at 1212 kw
Cruising speed: 250 km/hr
Hovering ceiling IGE: 3,170 m
Range: 780 km
Some models: 18 passengers and 2 crew members

2.4.5-b   Bell Model 206L-1, Lifebird, Felts Field Aviation

Two bladed rotor
1-X Rolls-Royce 250-C3OP turboshaft engine, 485 kw
Cruising speed: 215 km/hr
Range: 620 km
5 passengers and crew of two
Double door on port side for loading cargo

115

2.4.5-c Knoell Homes' Bell Model 222B

Two blade main rotor
2X Avco Lycoming LTS 101-750C-2 turboshaft engine 446 kw
Max speed: 155 mph
Service ceiling: 15,800 ft
Range: 434 miles
Eight passengers and two pilots

2.4.5-d Bell Model 406 CS test firing tow missile at Yuma Proving Grounds, Arizona

Four-blade composite rotor
Allison 250-C30R turboshaft engine, 650 shp

Cruising speed: 204 Km/h
Hovering ceiling: OGE: 3415 m
Range 550 km
Pilot and co-pilot/observer next to each other

2.5 Personal "Flying" Machines

"Flying" Machine	Description	Application
Rocket Back Pack	Rocket engine worn on back	Rapid deployment of personnel on earth.
Jet Back Pack	Jet engine worn on back	Rapid deployment of personnel on earth.
One Person Flying Platform	Engine mounted on platform. Pilot stands on platform.	Rapid deployment of personnel on earth. Predecessor to astronaut deployment machine for moon missions.
Two Person Flying Platform – POGO STICK	Engine mounted on platform. Two people stand on platform.	Rapid deployment of equipment and/or personnel on earth. Predecessor to astronaut deployment machine for moon missions.
Flying Chair	Engine mounted to chair. Pilot sits in chair.	Airplane pilot ejects over enemy territory and flies ejection seat to friendly territory.
Zero Gravity Belt	Rocket engine mounted on belt worn by pilot (astronaut).	Tested under zero-g conditions, during airplane dive, for simulation of astronaut maneuvering in space.

Cenkner Table

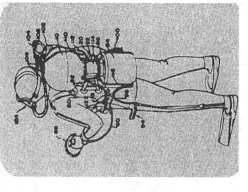

Underwater unit patent, employs closed cycle steam turbine, which also generates electrical power and oxygen

Lightweight pump fed machine, patent, with wings, to increase range

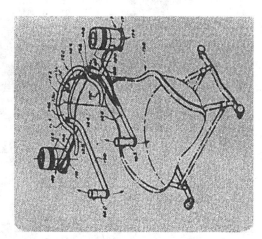

Flying chair patent

Reference

(1) *Small Rockets Find New Uses*, Rendezvous, Bell Aerosystems, Vol. V, No. 4, 1966.

116.5

2.5.1   Bell's Rocket Belt Back Pack

2.5.1-a   Bell's Data Sheet for Rocket Belt Back Pack

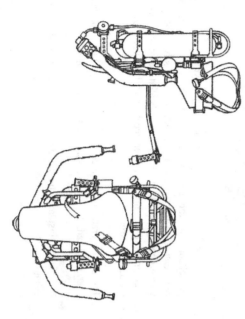

Propellant: (H_2O_2)	Hydrogen Peroxide
Propellant Weight:	47 lbs
Empty Weight:	63 lbs
Throttleable Thrust:	0-300 lbf
Maximum Range:	866 ft
Altitude:	80+ feet
Maximum Speed:	60+ mph
Record of Reliability:	100% reliability in more than 3,000 flights

BELL AEROSYSTEMS
BUFFALO, NEW YORK - A textron COMPANY

On April 20, 1961 a "Small Rocket Lift Device" was successfully flight tested. This unique system marked the beginning of a totally new dimension in the realm of flight. This device, known as the Bell Rocket Belt, was the first rocket powered system in the world which propelled man above the ground in controlled free flight.

The Bell Rocket Belt is a hydrogen peroxide propulsion system mounted on a fiberglass corset. Lift is provided by thrust from twin rocket nozzles which are fed by a central gas generator controlled by a throttle. Metal control tubes extend forward on each side of the operator. A control handle on one tube permits the operator to change his flight direction. A motorcycle-type hand throttle on the other tube allows him to regulate rocket thrust levels, thus controlling his rate of climb and descent. These controls permit complete freedom of flight, including forward, backward, sideward, up and down movement and even permits hovering in mid-air.

Since the Rocket Belt was first unveiled to the public at Fort Eustis, Virginia, in June 1961 more than 3,000 demonstrations have been conducted throughout the United States and Canada and on five continents.

The Rocket Belt concept presented by Textron's Bell Aerosystems has fostered new challenges for earth and space transportation, Present research and development efforts have successfully flight tested three new rocket propelled systems for transporting men and equipment on earth or over the surface of the moon. The new systems, extensions of the Rocket Belt concept, are a small Flying Chair and a standup version for transporting one or two men.

Although the present earth bound capabilities of the Bell Rocket Belt are limited to feasibility studies and public demonstrations, the freedom of movement and the restart properties of the Rocket Belt are an important feature for future lunar transportation applications.

117

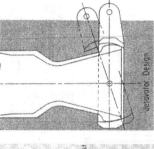

Jetavator Design

The direction of thrust and direction of travel were controlled by varying the plane of the exit nozzle.

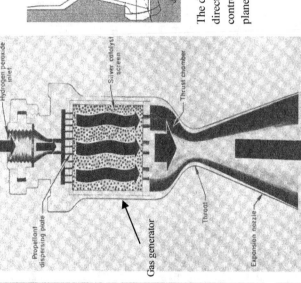

Hydrogen peroxide inlet

Silver catalyst screen

Thrust chamber

Propellant dispersing plate

Gas generator

Throat

Expansion nozzle

High pressure nitrogen was used to expel the hydrogen peroxide (H_2O_2) propellant from the 2 storage tanks, to the gas generator. A special plate dispersed the hydrogen peroxide across a silver catalyst screen, which decomposed it into 1300 F superheated steam and oxygen. This high pressure gas mixture is then accelerated through a nozzle to create the desired thrust.

The gas generator is mounted above the storage tanks and behind the operators head, thereby necessitating the use of a head heat shield.

In practice, insulated tubing diverts equal amounts of gas generator exhaust gas to two Jetavator nozzles which are located on each side of the operator, at hip level.

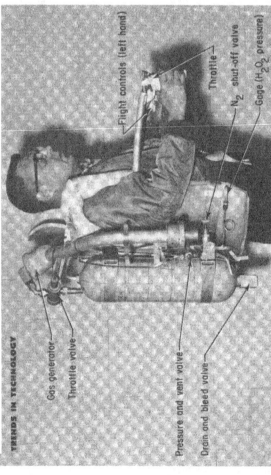

TRENDS IN TECHNOLOGY

Flight controls (left hand)

Throttle

N_2 shut-off valve

Gage (H_2O_2 pressure)

Gas generator

Throttle valve

Pressure and vent valve

Drain and bleed valve

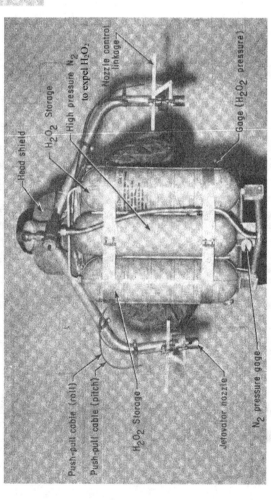

Head shield

H_2O_2 Storage

High pressure N_2 to expel H_2O_2

Nozzle control linkage

Gage (H_2O_2 pressure)

Push-pull cable (roll)

Push-pull cable (pitch)

H_2O_2 Storage

Jetavator nozzle

N_2 pressure gage

118

2.5.2 Bell's Jet Belt Back Pack

2.5.2-a Bell's Data Sheet for Jet Belt Back Pack

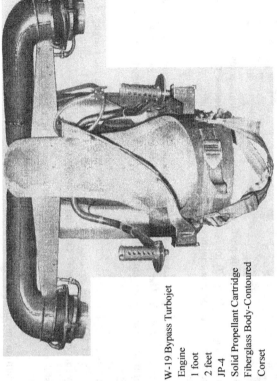

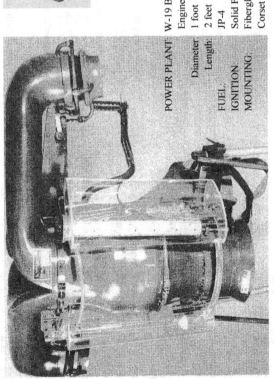

POWER PLANT:	W-19 Bypass Turbojet Engine
Diameter:	1 foot
Length:	2 feet
FUEL:	JP-4
IGNITION:	Solid Propellant Cartridge
MOUNTING:	Fiberglass Body-Contoured Corset

Developed under contract for the Department of Defense, Bell Aerosystems' Jet Flying Belt, brings a new dimension to individual mobility. Based on Bell experience with the highly successful rocket belt, this jet-powered version has a longer range and flight duration than the rocket belt, enabling the operator to perform a wide range of military missions and civilian tasks.

Power for the Jet Belt is supplied by a small bypass turbojet engine specifically designed for it. Featuring a high thrust-to-weight ratio and low fuel consumption, the engine throttling and flight maneuverability of the Jet Belt are easily controlled by motorcycle-type hand grips, giving various degrees of thrust and deflection to the nozzles. The entire personal propulsion unit is mounted on a fiberglass corset putting most of the weight comfortably on the operator's hips. The kerosene-type fuel which powers the unit is carried in clear plastic tanks which wrap around the engine, and is ignited by a solid propellant cartridge starting device, about the size of a shotgun shell.

With the Jet Flying Belt, the individual is capable of flying forward, backward, sideward, rotating on his vertical axis, executing coordinated turns and hovering.

Potential military applications for this new mobility system include flying over barbed-wire and mine fields, reconnaissance, counter-guerilla warfare, assault, perimeter guard and amphibious landings. Some civil applications are riot control, power line and pipeline patrols, photographic news coverage, rescue operations, traffic surveillance and microwave tower inspections.

A radio communications system is an integral part of the Bell Jet Flying Belt.

BELL AEROSYSTEMS
BUFFALO, NEW YORK – A textron COMPANY

119

2.5.3   POGO: Bell's Rocket Propelled Flying Platform

2.5.3-b   Proposed one man POGO flying platform, for transporting moon astronauts

One potential application of the flying platform was to transport astronauts across the surface of the moon, over a distance of 15 or 20 miles. Other potential moon missions, that were considered, were: rescue mission, transporting equipment, and transporting astronauts into orbit to rendezvous with a spacecraft.

With the moon POGO, the propulsion system is also mounted to a platform and the astronaut steps onto the platform to fly it. The proposed moon POGO would weigh about 150 lbs and be fabricated of titanium and aluminum. The platform at the front would be used to carry equipment or a passenger.

To evaluate this concept, Bell's one man POGO was tested by NASA at the Langley Research Center in Virginia. POGO was mounted to a tethered gimbal. The moon's 1/6 earth's gravity was simulated by supporting 5/6 of POGO's weight by a vacuum cylinder.

The successful tests showed that POGO was easier to operate in a simulated moon environment than on earth.

The moon platform was dropped when it was decided that a ground transport would be less risky.

2.5.3-a   Bell's two man POGO flying platform

As an extension to Bell's Rocket back pack technology, Bell built and tested three platform flying machines, nicknamed POGO (stick), that had the rocket propulsion systems mounted to the platform; the pilot would step onto the platform to operate it. Bell successfully built and tested three versions of the flying platform.

The first POGO was a one man unit, with a vertical post for the pilot to hold on to and with the propulsion system mounted in front of the pilot.

The second platform (above) was built for two men, with the propulsion unit situated between the pilot and the passenger.

The final platform was a one man version, with the propulsion unit mounted behind the pilot.

The units could be used for rapid deployment of personnel and equipment, and for rescue missions.

2.5.4 Bell's Emergency Ejection Seat, With Powered Parawing, for Combat Pilots

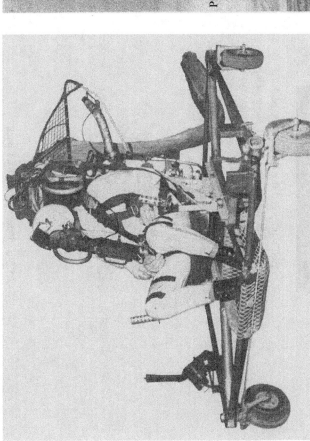

Labels on figure:
- Forward propelled motion
- Parasail
- Stored parachute for descent at flights end
- Crushable landing impact attenuator
- Turbojet engine
- Pilot ejection seat

2.5.4-a Prototype of Bell's emergency ejection seat with parasail for combat pilots

2.5.4-b Bell's proposal for emergency ejection seat with parasail for combat pilots

Under contract from the Air Force Flight Dynamics Laboratory, as part of the Air Force's Air Crew Escape Rescue System Capability (AERCAB) program, Bell Aero-systems was to design, build, and test a prototype (Figure 2.5.4-a) of Bell's proposed pilot rescue system; Figure 2.5.4-b.The proposed system is to weigh less than 600 pounds, produce speeds of more than 100 mph, with a range greater than 50 miles.

Bell engineers utilized a Bell Rocket Belt engine, to simulate the jet engine, during the prototype phase of the project, along with a modified T-33 aircraft seat.

A total of 24 helicopter drop tests were conducted. The folded parasail was attached to the wheeled simulated ejection seat and lifted to altitudes between 1,200 and 6,500 feet. After the unit was dropped, the parasail was deployed and the rocket engine was started.

Of the 24 drop tests, five were manned and powered, seven were manned but unpowered, and 12 were unmanned. With the hydrogen-peroxide rocket engine, 43 seconds of powered flight was obtained. A jet engine could operate for a longer period of time.

Reference: (1) Hole Card for Combat Pilots, Rendezvous, Bell Aerosystems, Vol VII, No. 4, 1969.
(2) Speth, R.F., Rust, J.L., AECAR Parawing Deployment and Articulation Flight Tests, National Technical Information Services, June 1971.

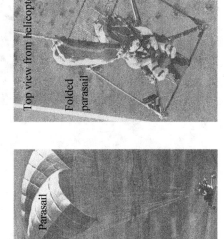

Top view from helicopter

Folded parasail

Parasail

2.5.4-c Helicopter drop test

2.5.5 Bell's 0-g Rocket Back Packs

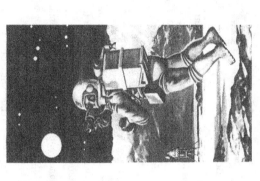

2.5.5-a   Bell's proposed 0-g rocket back pack for astronauts to travel outside spacecraft for making repairs, moving from one spacecraft to another etc.

2.5.5-c   Patent for 0-g back pack

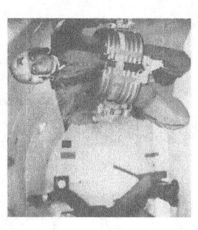

2.5.5-b   Bell's proposed 0-g rocket back pack for astronauts to travel on the moon

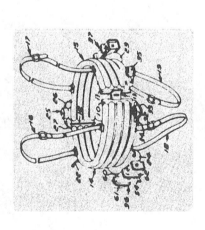

2.5.5-d   Testing 0-g rocket back pack under simulated space conditions by creating zero-gravity conditions in a diving airplane

122

2.6 Missiles

2.6-a History of the Development of Missile Technologies by the Bell-aerospace-company

Name	Type	Range	Speed	Ceiling	Propulsion	Bell Contribution
CT-41 (PQM-56)	Target missile	14 minutes	Mach 3.1	65,000 ft	First stage: 2 solid propellant rocket engines Second stage: 2 ramjets	Test, evaluate and manufacture the missile for the U.S. Navy.
GAM-63 Rascal	Air-to-Surface, nuclear warhead	100 miles	Mach 3.5	70,000 ft	Bell XLR-67-BA-1 liquid propellant rocket engine	Construct missile. Develop rocket engine and bomb fusing system. Joint development of guidance system with RCA and Texas Instrumentation.
LGM 30G Minuteman III	Surface-to-Surface, nuclear warhead	8,100 miles	15,000 mph	700 miles	3 Solid propellant rocket engine stages 1 Post boost liquid propellant rocket engine	Develop and manufacture post boost liquid propellant rocket engine.
Sentry Interceptor	Surface-to-Air kinetic energy weapon	Not available	Not available	Not available	Booster: solid propellant rocket engine Terminal control system: liquid propellant rocket engine Control system..	Liquid propellant rocket engines for terminal control of the missile.
United Kingdom HAS A3TK	Submarine launched ballistic missile	Not available	Not available	Not available	First stage: unknown Post boost: liquid propellant rocket engine	Develop and manufacture post boost liquid propellant rocket engine.
RTV-A-4/X-9 Shrike	Air-to-Surface, test bed for Rascal	50 miles	>Mach 1.5	65,000 ft	Bell XLR65-BA-1 liquid propellant rocket engine	Develop missile.
MIT/Bell AAM-N-5 Meteor	Air-to-Air blast fragmentation	25 miles	>Mach 2	Not available	First stage: solid propellant rocket engine Second stage : liquid propellant rocket engine	Develop and build the air frame.
Kinetic Energy Weapon (KEW)	Ram target with non explosive payload	Not available	Not available	Not available	Bell N_2H_4/N_2O_4 rocket engine	Develop missile
Nike	Surface to Air anti aircraft fragmentation	N.A.	N.A.	N.A.	N.A.	N.A.
Kingfish	Torpedo	N.A.	N.A.	N.A.	N.A.	Rocket engine
MX missile Peacekeeper	Surface-to-Surface, nuclear warhead (MIRV)	Not applicable	Not applicable	Not applicable	Not applicable	Design and develop shock isolation system to protect MX missile launch vehicle from shock and vibration. Develop diaphragm tank for stage four.

Cenkner Table

References

(1) *RASCAL Development Engineering Inspection*, Bell Aircraft Corporation, Report No. 66-945-010, April 1957.

(2) *Bell to Test CT-41 for Navy*, Rendezvous, Bell Aerosystems , Vol 1, No. 4.

(3) *Post Boost Propulsion Subsystem for Minuteman III*, Bell Aerospace Company, Dec, 1969.

(4) *Kinetic Energy Weapon Program Divert Propulsion Technology*, Bell Aerospace Textron, Report No. 8905-927012. March 14, 1986.

(5) *Kinetic Energy Weapon (KEW) Propulsion Technology*, Bell Aerospace Textron, Report No. 8905-927014, March 26, 1986.

(6) Ira G. Ross Aerospace Museum, Buffalo, N.Y.

122.5

2.6.1   Bell Flight Testing of French Supersonic CT-41 Recoverable Target Missile

The CT-41 missile was being evaluated for possible use, as a target missile, in the development of air-to-air and ground-to-air missile defense systems.

The missile is launched, from the ground, by two solid propellant boosters and powered in flight by two wing-tip ram jets, which can achieve a maximum speed of Mach 3.1 at an altitude of 65,000 feet. In flight, the missile is radio controlled; it can be commanded from the ground to turn left or right, climb, dive, change speed, or parachute down for recovery and reuse. All missile sections are waterproof. They will continue to float until picked up by a recovery vehicle.

The missile can simulate larger bomber or fighter aircraft by using onboard electronic radar augmentation equipment.

The missile is about 34 feet long, has a wingspan that is close to 12 feet, and weighs 6,650 pounds at launch. The solid propellant rocket engines accelerate the missile to Mach 1.7. At this speed, the ramjets can be started to further accelerate the missile to Mach 3.1.

Bell purchased six CT-41 missiles, under Navy contract, to conduct a two phase flight evaluation program, with each phase consisting of six flights.

Phase I: Determine its compatibility with the Pacific Missile Range launching, ground control and tracking systems. In addition, determine if it can meet the advanced performance needs of the Navy.

Phase II: Evaluate its acceptability as a target for the development of air-to-air and surface-to-air missiles. Onboard equipment would determine how close attack missiles approach this target missile.

The tests were successfully completed; the missile met Navy requirements.

Bell obtained a production license, in March 1960, from Nord Aviation, to manufacture missiles for the Navy. The missiles were retired by the Navy in 1971.

Reference

(1)   *Bell to Test CT-41 for Navy*, Rendezvous, Bell Aerosystems, Vol.1, No. 4, Nov/Dec, 1962.

(2)   *CT-41 – A Supersonic Bull's Eye*, Rendezvous, Bell Aerosystems, Vol II, No. 4, Aug 1963.

Solid propellant booster

Wingtip ram jets

2.6.1-a   Nord Aviation, of France, developed the CT-41 supersonic target missile

2.6.2 Bell's GAM-63 Rascal Guided Missile

2.6.2-a Bell's Data Sheet for the GAM-63 Rascal Missile -- a guided rocket powered supersonic air-to-ground nuclear armed missile

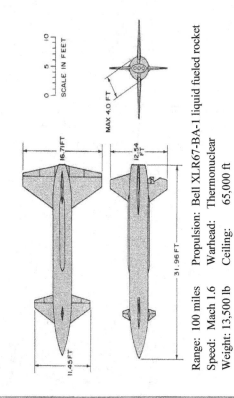

16.71FT

12.54 FT

31.96 FT

11.45FT

MAX 4.0 FT

SCALE IN FEET

Range: 100 miles
Speed: Mach 1.6
Weight: 13,500 lb

Propulsion: Bell XLR67-BA-1 liquid fueled rocket
Warhead: Thermonuclear
Ceiling: 65,000 ft

BELL RASCAL

The Air Force RASCAL is an air-to-ground guided missile designed to be carried aloft by B-47 bombers of the Strategic Air Command. It is released at high altitude, miles from its target, so that bomber and crew are not exposed to local defenses. The rocket-powered missile actually flies toward the target at supersonic speed while the launching airplane has turned and is enroute to its base. This has earned the missile, designed and built by Bell Aircraft Corporation, the nickname "Crew Saver." It is capable of carrying an atomic or hydrogen warhead.

BELL *Aircraft* CORPORATION
POST OFFICE BOX ONE · BUFFALO 5. NEW YORK

2.6.2-b Geometry of GAM-63 RASCAL guided missile. The first successful air launch was achieved in July 1955, while the first production missile was delivered to the USAF on October 30, 1957. The RASCAL program was terminated on September 28, 1958.

Six RASCAL Systems

- Guide the missile to the target
- Control the missile during the terminal phase
- Internal electrical power and distribution
- Hydraulic system to move flight control surfaces and antennas
- Propulsion system
- Fusing system to ignite the warhead at ground impact or at a predetermined altitude

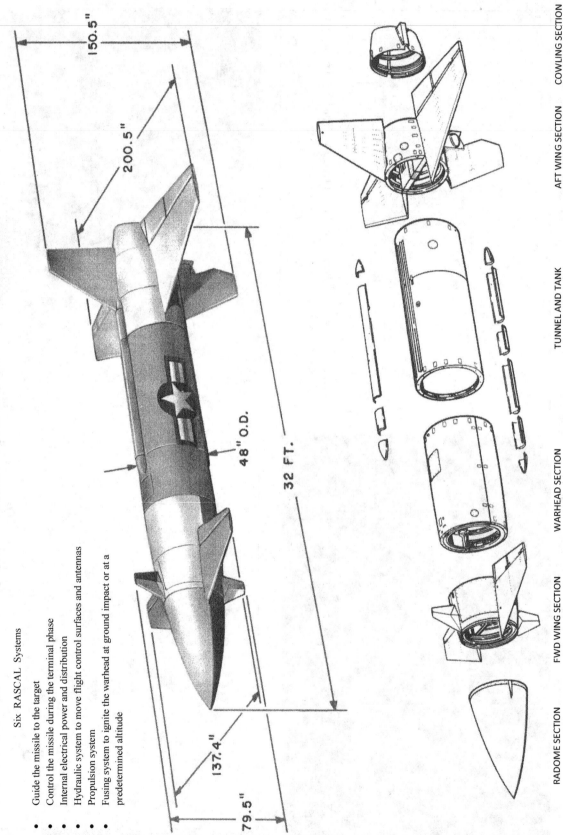

RADOME SECTION   FWD WING SECTION   WARHEAD SECTION   TUNNEL AND TANK   AFT WING SECTION   COWLING SECTION

2.6.2-c Cut-Away views of GAM-63 RASCAL guided missile

Six RASCAL Systems

• Guide the missile to the target by bombardier
• Control the missile during the terminal phase
• Internal electrical power and distribution
• Hydraulic system to move flight control surfaces and antennas
• Propulsion system
• Fusing system to ignite the warhead at ground
  impact or at a predetermined altitude

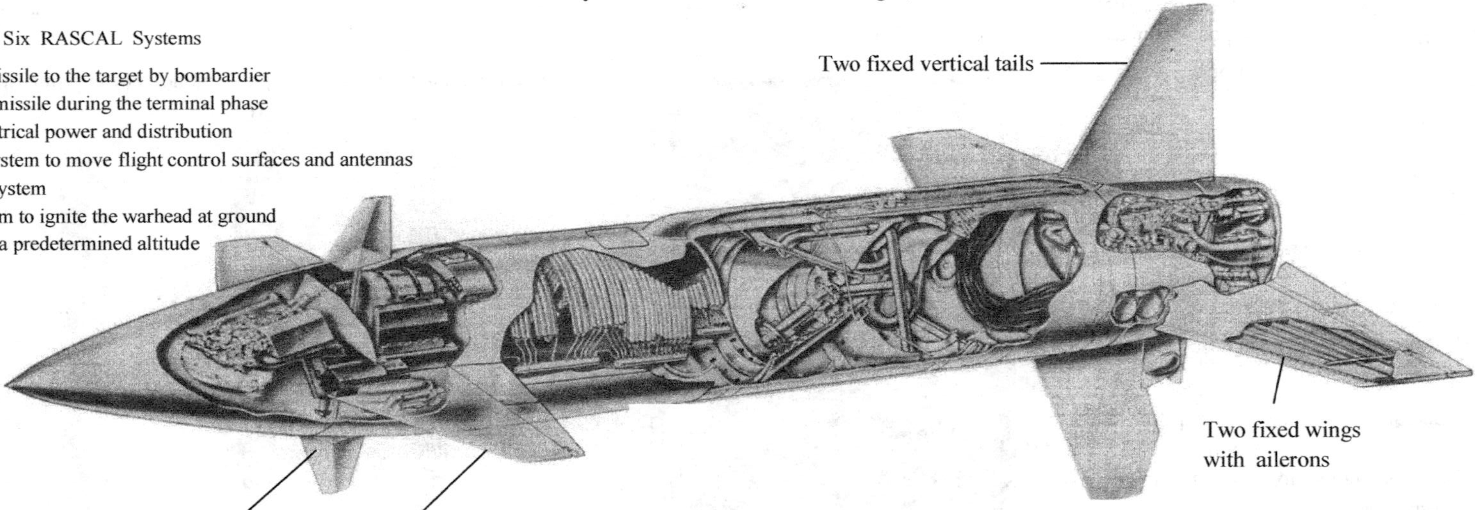

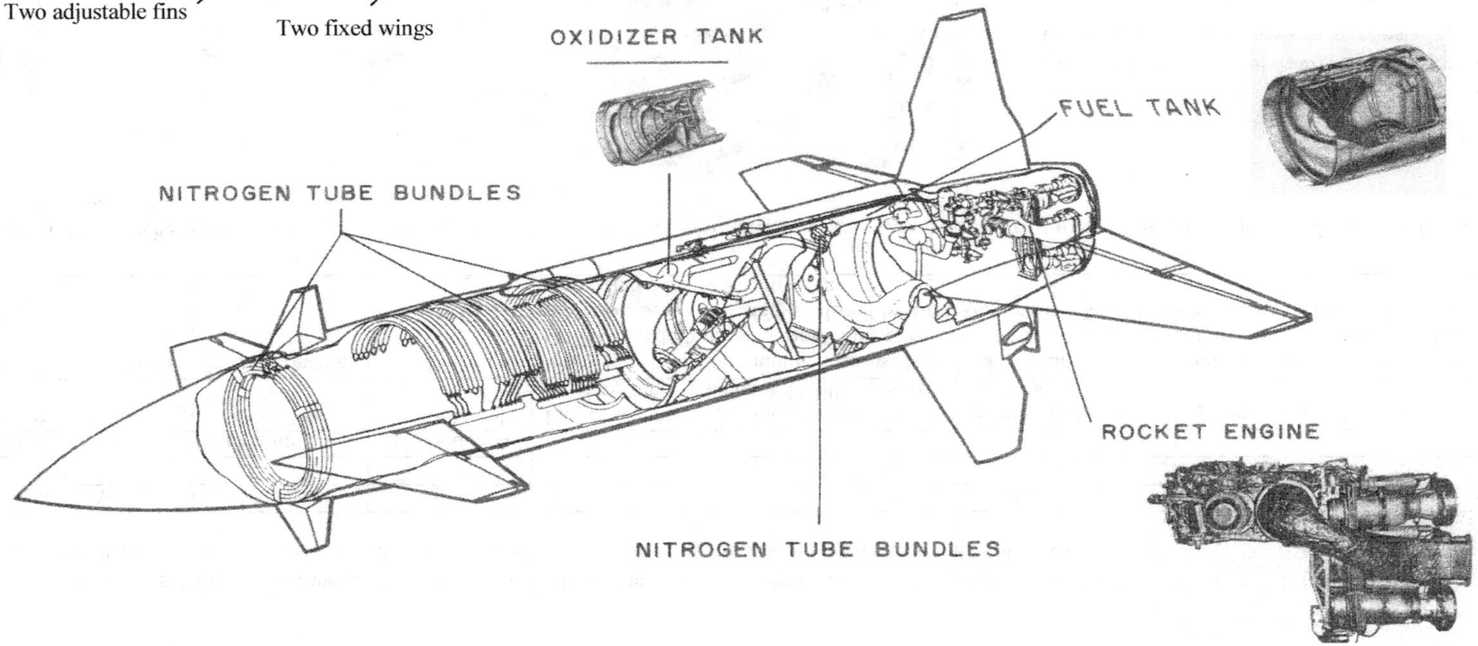

Two fixed vertical tails

Two fixed wings
with ailerons

Two fixed wings

Two adjustable fins

FUEL TANK

OXIDIZER TANK

ROCKET ENGINE

NITROGEN TUBE BUNDLES

NITROGEN TUBE BUNDLES

126

2.6.3 Bell's Post Boost Rocket Propulsion System for LGM 30G Minuteman III ICBM Missile

Minuteman is a three stage solid propellant intercontinental ballistic missile (ICBM) that carries a nuclear warhead. The land based missiles are stored in hardened underground silos, on ready alert, for immediate retaliation if the U.S. is attacked. The original missile design has been upgraded a number of times.

Model	Length	Weight	Range	Speed	Upgrades
Minuteman I	55.9 ft	65,000 lb	Intercontinental	>15,000 mph	
Minuteman II	59.8ft	70,000 lb	Intercontinental	>15,000 mph	Larger second stage motor, improved guidance, greater range, greater payload, more flexible targeting, and an increased ability to survive an attack.
Minuteman III	59.8ft	76,000 lb	Intercontinental	>15,000 mph	Improved third stage motor, increased payload, Multiple Independently Targeted Reentry Vehicles (MIRV). Addition of a **post boost attitude and velocity control propulsion system (PBPS)** designed, developed, and produced by **Bell Aerospace**. Decoys, warheads, and radar confusing chaff are included in the re-entry vehicle.

The first Minuteman I was launched on February 1, 1961. On September 24, 1961 the first Minuteman II was launched. The first research and development Minuteman III was launched on April 11, 1969.

The Triad – the national deterrent force concept – is comprised of three weapons systems:

(1) Minuteman land based missiles
(2) Strategic Air Command's long range bombers
(3) U.S. Navy's nuclear submarines

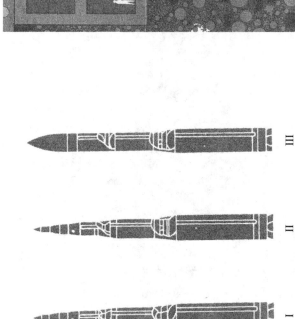

Launch silos:
  80 feet underground
  12 feet in diameter
  80 ton concrete slab, riding on rails and covering the missile

I   II   III

All missiles are roughly six feet in diameter, at widest point.

Reference

(1) Minuteman ….The Dynamic Deterrent, Rendezvous, Bell Aerospace, Vol. X / Spring 1971.

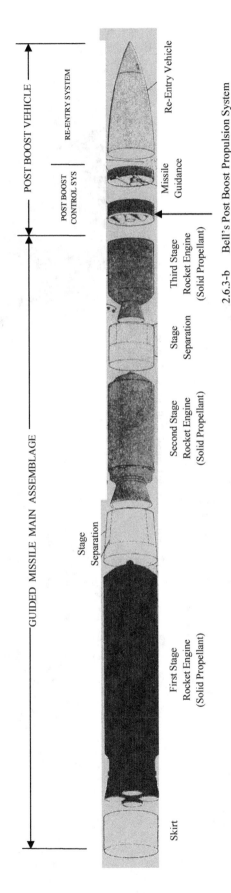

2.6.3-a   Components of Minuteman III Missile

2.6.3-b   Bell's Post Boost Propulsion System

The LGM 30G Minuteman III ICBM is a silo launched (L) guided missile (M) designed to attack ground targets (G).

Model 30G has three solid propellant stages and a new liquid propellant fourth stage, or post boost propulsion system (PBPS), developed and built by the Bell Aerospace Company. The PBPS was added to increase the range and maneuverability of the missile, and to increase the payload; the range was increased to over 8,000 miles. Other modifications include upgraded electronics to improve accuracy, and to reduce vulnerability in a nuclear environment. This model can accommodate up to three independently targeted nuclear warheads.

The PBPS is designed to supply axial and translational thrust, in addition to attitude control torques, during the final phase of the trajectory.

The PBPS must function reliably after up to five (and possibly ten) years of storage, for up to eight minutes.

Reference

(1) POST BOOST PROPULSION SUBSYSTEM for MINUTEMAN III,
    Bell Aerospace Company

128

2.6.3-c   Details on Bell's Propulsion System Rocket Engine (PSRE)

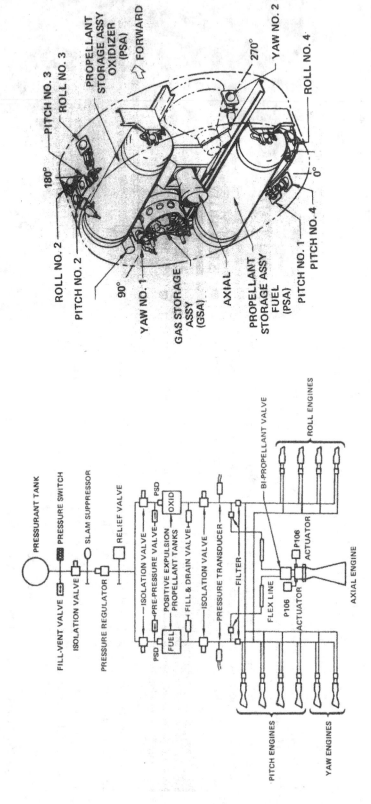

PITCH NO. 3
ROLL NO. 3
PROPELLANT STORAGE ASSY OXIDIZER (PSA)
FORWARD
180°
ROLL NO. 2
PITCH NO. 2
90°
270°
YAW NO. 2
ROLL NO. 4
YAW NO. 1
GAS STORAGE ASSY (GSA)
AXIAL
PROPELLANT STORAGE ASSY FUEL (PSA)
PITCH NO. 1
PITCH NO. 4
0°

2.6.3-d   Attitude Control (Roll) Rocket Engine

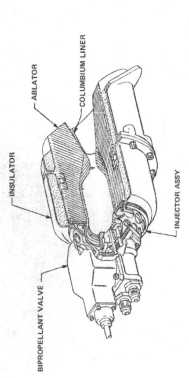

ABLATOR
COLUMBIUM LINER
INSULATOR
INJECTOR ASSY
BIPROPELLANT VALVE

PRESSURANT TANK
PRESSURE SWITCH
SLAM SUPPRESSOR
RELIEF VALVE
PSD
OXID
BI-PROPELLANT VALVE
ROLL ENGINES
FILL-VENT VALVE
ISOLATION VALVE
ISOLATION VALVE
PRE-PRESSURE VALVE
ACTUATOR
P106
PRESSURE REGULATOR
PSD
POSITIVE EXPULSION PROPELLANT TANKS
FUEL
FILL & DRAIN VALVE
ISOLATION VALVE
PRESSURE TRANSDUCER
FILTER
FLEX LINE
P106
ACTUATOR
AXIAL ENGINE
PITCH ENGINES
YAW ENGINES

129

2.6.3-e Axial rocket engine with direct torque motor actuated bipropellant valve, a fixed orifice injector, a beryllium thrust chamber, a thermal insulation jacket, and a gimbal ring engine mount.

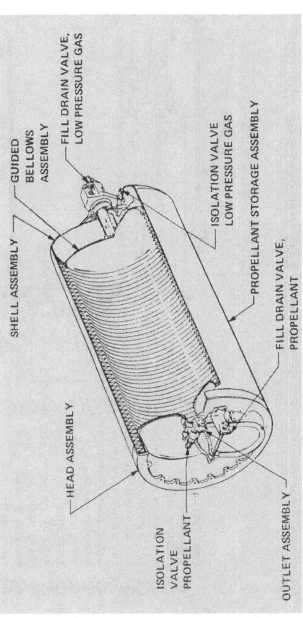

2.6.3-f Propellant storage tank assembly. One assembly was used for each propellant. Each unit contains a positive expulsion bellows made from 0.007 inch thick type 347 stainless steel.

SHELL ASSEMBLY

GUIDED BELLOWS ASSEMBLY

FILL DRAIN VALVE, LOW PRESSURE GAS

ISOLATION VALVE LOW PRESSURE GAS

PROPELLANT STORAGE ASSEMBLY

HEAD ASSEMBLY

FILL DRAIN VALVE, PROPELLANT

ISOLATION VALVE PROPELLANT

OUTLET ASSEMBLY

2.6.3-g PSRE system schematic

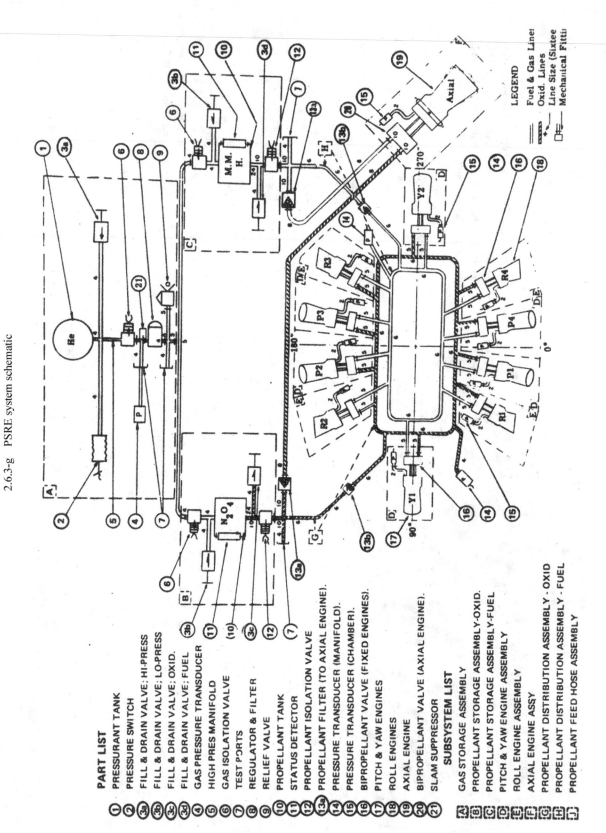

PART LIST

1. PRESSURANT TANK
2. PRESSURE SWITCH
3a. FILL & DRAIN VALVE: HI-PRESS
3b. FILL & DRAIN VALVE: LO-PRESS
3c. FILL & DRAIN VALVE: OXID.
3d. FILL & DRAIN VALVE: FUEL
4. GAS PRESSURE TRANSDUCER
5. HIGH PRES MANIFOLD
6. GAS ISOLATION VALVE
7. TEST PORTS
8. REGULATOR & FILTER
9. RELIEF VALVE
10. PROPELLANT TANK
11. STATUS DETECTOR
12. PROPELLANT ISOLATION VALVE
13a. PROPELLANT FILTER (TO AXIAL ENGINE).
14. PRESSURE TRANSDUCER (MANIFOLD).
15. PRESSURE TRANSDUCER (CHAMBER).
16. BIPROPELLANT VALVE (FIXED ENGINES).
17. PITCH & YAW ENGINES
18. ROLL ENGINES
19. AXIAL ENGINE
20. BIPROPELLANT VALVE (AXIAL ENGINE).
21. SLAM SUPPRESSOR

SUBSYSTEM LIST

A. GAS STORAGE ASSEMBLY
B. PROPELLANT STORAGE ASSEMBLY-OXID.
C. PROPELLANT STORAGE ASSEMBLY-FUEL
D. PITCH & YAW ENGINE ASSEMBLY
E. ROLL ENGINE ASSEMBLY
F. AXIAL ENGINE ASSY
G. PROPELLANT DISTRIBUTION ASSEMBLY - OXID
H. PROPELLANT DISTRIBUTION ASSEMBLY - FUEL
J. PROPELLANT FEED HOSE ASSEMBLY

131

2.6.3-h   Minuteman III PSRE critical system requirements and status as of March 26, 1986

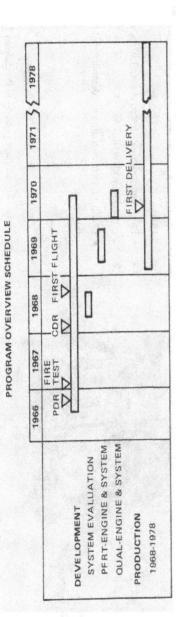

PROGRAM OVERVIEW SCHEDULE

Major Components

- Gas Storage Assembly for Helium.

- Propellant Storage Assembly with Bellows Expulsion for
  Nitrogentetroxide and Monomethylhydrazine .

- 10 Pitch, Yaw and Roll Control Engines

- Axial Engine

- Associated Actuation, Feed and Control Components

- All Operational Equipment Mounted in a MM III Section 52
  Inches in Diameter, 19 Inches Deep.

Minuteman III PSRE Critical System Requirements

- Nuclear Hardened System.

- Loaded at the Factory, Highway Transportable.

- Long Term Leak Free Storage.

- Safety of Personnel and Equipment.

- Minimum Maintenance.

- Instant Readiness.

Minuteman III PSRE Achievement as of March 26, 1986

- 867 Production Deliveries.

- Zero System Maintenance Subsequent to Delivery.

- Over 2.8 Million Engine Firings.

- Over 61 Million Hours MTBF.

- 10,324 Years of Cumulative Leak-Free Zero Maintenance.

- 1,358,890 Miles Without Leakage or Damage.

- 74 Ground Tests.

- Surveillance Firings of 3, 4, 5, 6, 7 and 15-Year Old Systems (5 Each).

- 12 Successful Flight Tests After 10 to 15 Years Storage.

- System Life –
  - Original Requirement : 5 Years
  - Original Goal: 10 Years
  - Expected to Exceed: 30 Years

- USAF Official Evaluation "100% Demonstrated System Reliability"

- All 174 Flight Tests Completed Successfully.

132

2.6.4 Bell's Sentry Interceptor Reaction Control System

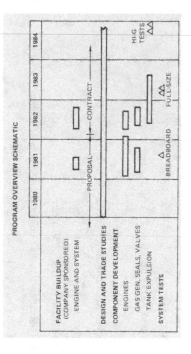

Sentry Interceptor – A ground launched missile for ballistic missile defense against incoming missiles, low endoatmospheric altitudes, at end-of-flight of incoming missile

Application – Responsible for terminal maneuvers in a severe nuclear environment

System – Includes a solid propellant gas generator that supplies gas to pressurize the liquid propellant tanks, in addition to providing roll control

System Design – Packaged around the first stage solid boost motor nozzle

Propellants – Piston tanks are used to store the tetrozide and monomethyl hydrazine propellants

Rocket Engines – Pitch and Yaw engines are ablative lined. To optimize thrust amplification through jet interaction with the external flow field, a slotted exit nozzle was incorporated.

Status – Program terminated 1984, when the government decided not to deploy this type of interceptor

Sentry Achievements as of 1984

- System activation time demonstrated with full size control system

- Light weight components: ~9000 pound thrust engine weighing 10 pounds

- Turn flow ablative engines met requirements:

  Performance

  Rapid Response

  Deep Throttling Durability

- Rapid expulsion of live propellants by hot gas with full size tanks

- Propellant tank shock tested at over 130 g's while undergoing hot gas expulsion, with no leakage

- Control system operating life demonstrated for worst case mission duty cycle

Reference

(1) *Kinetic Energy Weapon Propulsion Technology*, Report No. 8905-927014, Bell Aerospace Textron, March,26,1986

2.6.4-a   Second stage liquid propellant rocket engines package around first stage exhaust nozzle

133

2.6.5 Post Boost Propulsion System Developed by Bell for the United Kingdom

HAS – A liquid monopropellant Hydrazine Actuation System developed for the U.K.

Application – Supplies post boost propulsion for accurate delivery of ballistic missiles that were launched from a submarine (A3TK SLBM)

System – One hot gas subsystem for pressurizing the axial thruster propellant tanks and one hot gas subsystem for a monopropellant reaction control thrusters

System Design   -- Two hydrazine hot gas pressurization systems

     -- Propellant storage tanks incorporating a reversing diaphragm for propellant expulsion

     -- Three rocket engine clusters. Each cluster consists of four pitch, yaw and roll thrusters

Delivery – Propellant tanks were prepackaged at the factory and shipped to the U.K.

Installation – Bell designed and built a turn-key facility for use with the hardware

Tests -- To demonstrate the required 10 year life, thermally accelerated compatibility tests were executed

     -- A propellant leak detector was included

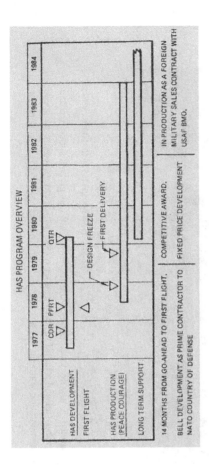

HAS Achievements as of March 26, 1986

- First flight 14 months after go-ahead
- 120 production deliveries
- Over 11 million firings of the pitch/yaw (45 lbf) and roll (9 lbf) engines
- Zero system maintenance after delivery
- Five year storage demonstrated and continuing
- 1254 tank-years of leak-free zero maintenance
- 16 systems ground tests
- Perfect safety record
- All 20 flight tests (submarine launched) completed successfully

Reference

(1) *Kinetic Energy Weapon Propulsion Technology*, Report No. 8905-927014, Bell Aerospace Textron, March,26,1986

134

2.6.5-a   HAS Post Boost Propulsion System

2.6.6   Shrike and Meteor Guided Missiles

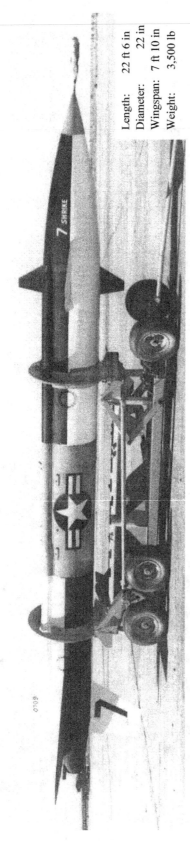

Length: 22 ft 6 in
Diameter: 22 in
Wingspan: 7 ft 10 in
Weight: 3,500 lb

2.6.6-a   Bell RTV-A-4/X-9 Shrike guided missile was a reduced size version of the Rascal air-to-surface missile. It was used to test the Rascal aerodynamic design, the propulsion and radio control systems, and handling procedures. It was powered by a Bell XLR65-BA-1 liquid propellant rocket engine. It had a range of 50 miles, a ceiling of 65,000 feet, and a top speed of more than Mach 1.5. In November 1950, it was successfully flown.

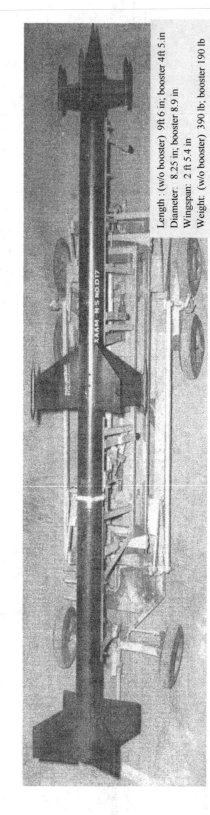

Length : (w/o booster) 9ft 6 in; booster 4ft 5 in
Diameter: 8.25 in; booster 8.9 in
Wingspan: 2 ft 5.4 in
Weight: (w/o booster) 390 lb; booster 190 lb

2.6.6-b   MIT/Bell AAM-N-5 Meteor air-to-air guided missile. MIT was awarded a missile development contract, while Bell was given a subcontract to develop and build the airframe. The missile utilized a solid propellant booster and a liquid fueled sustainer rocket, which gave a maximum speed greater than Mach 2, with a range of 25 miles. It carried a 25 lb fragmentation warhead. Test launches were initiated in 1951 but the contract was cancelled in 1953.

2.6.7 Bell's Kinetic Energy Weapon Propulsion System

KEW – Kinetic energy weapon that employs a missile, with a non-explosive payload, to ram a target (satellite or missile) at high speed and high kinetic energy, to destroy the target.

Attitude Control Thrusters – Used to control roll, pitch, and yaw of the missile.

Divert Thrusters – Used to translate the missile perpendicular to the flight path.

Operation of Attitude Control Thruster – When an electrical signal is sent to the electromagnet, the valve opens and the high pressure gas flows out the nozzle, producing the desired thrust.

Positive Expulsion Propellant Tanks – When warm gas is applied to the rolling diaphragm, it expands and pushes the propellant out of the tank (see section on storage tanks).

SEEKER PLUS G&C

PROPELLANT
- N_2O_4/N_2H_4

PROPELLANT TANKS
- CONTOURED ROLLING DIAPHRAGM

PRESSURIZATION
- SELF PRESSURIZING GAS DRIVEN INTENSIFIER

CONTROL
- 3 AXIS STABILIZED

DIVERT THRUSTERS
- 300LB THRUST (4)
- BIPROPELLANT

ATTITUDE CONTROL THRUSTERS
- BILEVEL 10 & 1.7 LBF (8)
- WARM GAS

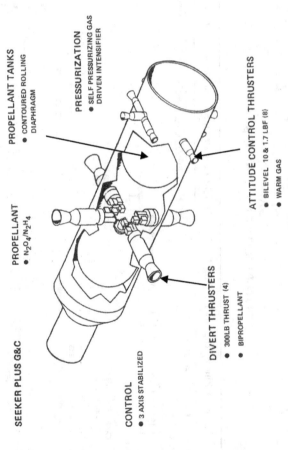

2.6.7-a   KEW propulsion system

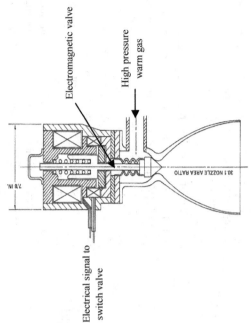

Electromagnetic valve

High pressure warm gas

7/8 IN.

30:1 NOZZLE AREA RATIO

Electrical signal to switch valve

2.6.7-c   Bi level warm gas attitude control engine

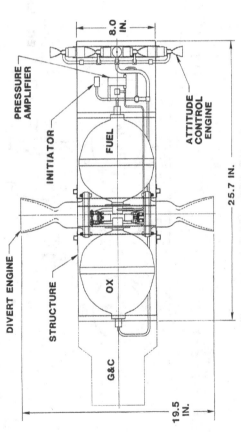

PRESSURE AMPLIFIER

INITIATOR

FUEL

ATTITUDE CONTROL ENGINE

DIVERT ENGINE

STRUCTURE

OX

G&C

8.0 IN.

25.7 IN.

19.5 IN.

2.6.7-b   KEW propulsion system layout

2.6.7-d Kinetic Energy System schematic

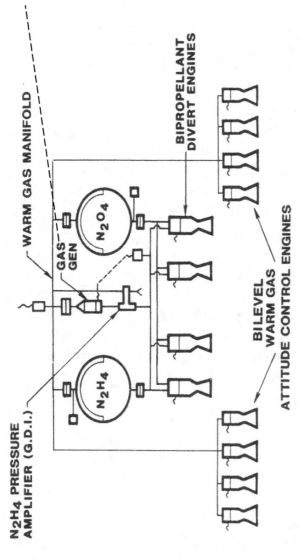

N₂H₄ PRESSURE AMPLIFIER (G.D.I.)
WARM GAS MANIFOLD
GAS GEN
N₂O₄
N₂H₄
BIPROPELLANT DIVERT ENGINES
BILEVEL WARM GAS ATTITUDE CONTROL ENGINES

Gas Generator – Generates warm gas for use with attitude control rockets and for extending propellant storage tank bellows to push propellant from tanks (see section on storage tanks).

2.6.7-f   Status of Critical Technologies

(1) Performance & Durability of Divert Thruster

- Mature high performance injector design successfully demonstrated in aluminum during Sentry program.
- Pilot operated fast response bipropellant valve is lightweight derivative of valve successfully demonstrated on Sentry program.

(2) Performance & Operation of Aluminum Rolling Diaphragm Propellant Tanks During Flight Dynamic Environment and Propellant Delivery.

- Similar aluminum diaphragm tanks successfully developed and demonstrated.
- Tank and Diaphragm integrity demonstrated during shock and random vibration tests.

(3) Operation & Performance of Gas Driven Intensifier Warm Gas Pressurization System.

- Successfully demonstrated by Rocket Research Department under contract to AFRPL.
- Successfully demonstrated during nine propellant expulsion tests.

References

(1)-*Kinetic Energy Weapon Program – Divert Propulsion Technology*, Report No. 8905-927012, Bell Aerospace TEXTRON, 14 March 1986

(2)-*Kinetic Energy Weapon (KEW) Propulsion Technology*, Report No. 8905-927014, Bell Aerospace TEXTRON, March 26, 1986.

Four Phase Program Leading To Flight Weight Propulsion System Demonstration

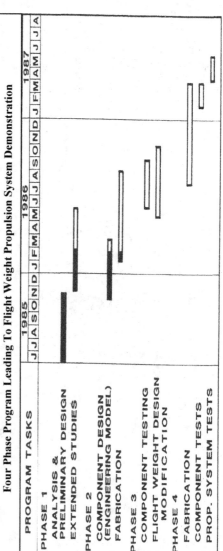

PROGRAM TASKS	1985	1986	1987
	J J A S O N D	J F M A M J J A S O N D	J F M A M J J A
PHASE 1			
ANALYSIS & PRELIMINARY DESIGN			
EXTENDED STUDIES			
PHASE 2			
COMPONENT DESIGN (ENGINEERING MODEL)			
FABRICATION			
PHASE 3			
COMPONENT TESTING			
FLIGHT WEIGHT DESIGN MODIFICATION			
PHASE 4			
FABRICATION			
COMPONENT TESTS			
PROP. SYSTEM TESTS			

2.6.7-e Four Phase Program

2.6.7-g Altitude vacuum chamber for testing propulsion system of kinetic energy weapon

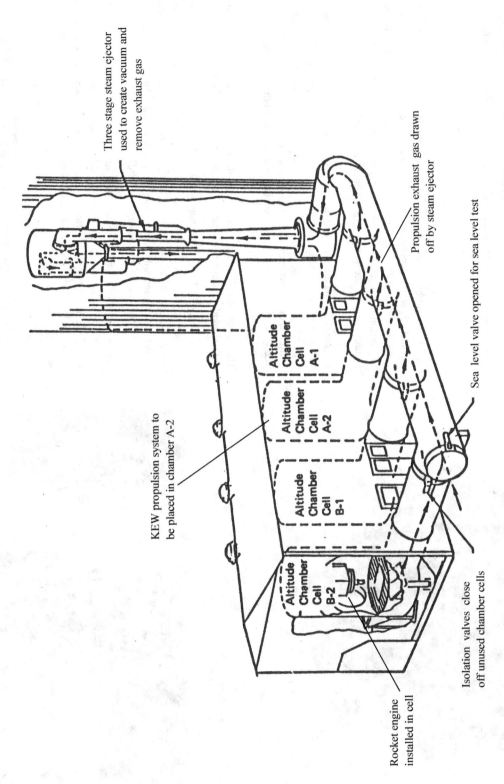

Three stage steam ejector used to create vacuum and remove exhaust gas

Propulsion exhaust gas drawn off by steam ejector

KEW propulsion system to be placed in chamber A-2

Altitude Chamber Cell A-1

Altitude Chamber Cell A-2

Altitude Chamber Cell B-1

Altitude Chamber Cell B-2

Sea level valve opened for sea level test

Isolation valves close off unused chamber cells

Rocket engine installed in cell

Altitude test chamber A-2 is ten feet in diameter and sixteen feet high. The three stage steam ejector system can remove up to 1400 lb per hour of noncondensible exhaust gas while maintaining a vacuum that is equivalent to an altitude of 120,000 feet. The exhaust gas can be effectively scrubbed. All test data is recorded in the Data Center.

138

2.6.8 Bell's Extendible Nozzle Cone

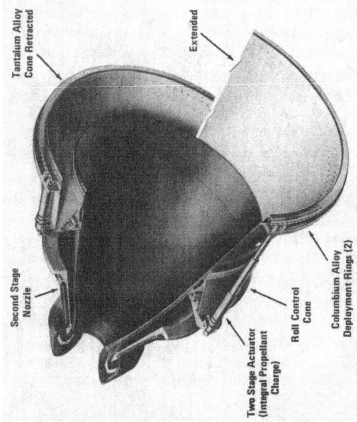

Tantalum Alloy Cone Retracted

Second Stage Nozzle

Two Stage Actuator (Integral Propellant Charge)

Roll Control Cone

Columbium Alloy Deployment Rings (2)

Extended

2.6.8-a Bell's Data sheet on Extendible Nozzle Cone that reduces the installed length to 1/3 of the extended length.

Bell's Extendible Cone consists of a convoluted nozzle (U.S. Patent No. 3,711,027) with a deployment system of 3 or more pneumatic actuators. The cone is formed with a portion folded back on itself (convoluted), reducing the installed length to 1/3 of the deployed operating configuration. The actuators extend the nozzle by a simple roll-through.

Engine performance and maximum payload are foremost priorities in space. Bell's Extendible Exit Cone offers a low-cost approach to optimum Operating efficiency. Up to 7% increased engine performance can be achieved or the space saved by the retracted cone can increase fuel capacity. Range and performance of existing and future engines will be improved significantly.

The Extendible Exit Cone is a unit of uniform thickness with no longitudinal welds; the cone is shear-spun from a rolled flat sheet.

An Extendible Exit Cone was successfully tested at the USAF Arnold Engineering Development Center as part of the USN Trident engineering development phase. A 30-inch diameter nozzle, fabricated from a ductile tantalum/tungsten alloy was fired in a high altitude chamber (100,000 ft) fully demonstrating this concepts feasibility.

Bell's Extendible Exit Cone can add hundreds of miles to the range of a ballistic missile or add the potential of greater payload capacity. It is adaptable to, and equally efficient for, all types of rocket engines, solid or liquid-fueled. Bell's Extendible Exit Cone is ready for specific operational applications.

Bell Aerospace TEXTRON

Division of Textron Inc.

2.7 Bombs

2.7.1 Bell's VB-13/ASM-A-1 Guided Vertical Bomb

Weight:	12,000 lb
Length:	21 ft
Diameter of body:	3 ft 2 in
Diameter of circular shroud:	4 ft 6 in
Propulsion:	None
Warhead:	6,700 lb high explosive
Control:	Radio control from drop plane

Gulf Research Corporation developed a prototype of the Tarzan Bomb. Bell received a contract to build ten more bombs, in mid 1945.

The bomb would free fall from the drop plane. After release, a flare in the tail of the bomb was ignited so the bombardier could trace the trajectory of the bomb. If it deviated from the desired trajectory, the bombardier would use the radio control system to adjust the movable tail shroud to get the bomb back on target. The range and azimuth of the bomb could be controlled.

The bomb had a radio controlled rudder and elevator, and four ailerons.

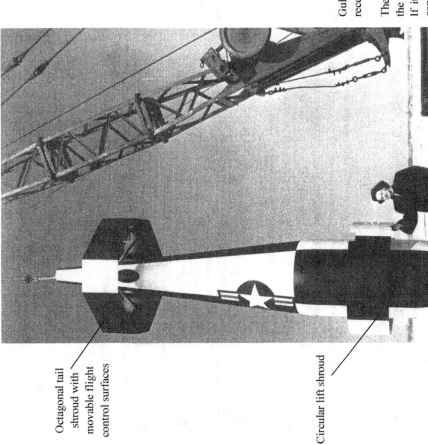

Octagonal tail shroud with movable flight control surfaces

Circular lift shroud

2.7-a YASM-A-1 Vertical Bomb   USAF photo

140

2.8 Radar Landing Systems for Airplanes

2.8.1 Development of Bell's Microwave Landing Systems

Because of the high incidence of accidents, during landings in adverse conditions the Navy was extremely interested in obtaining a system that would automatically land any aircraft on any aircraft carrier.

The ideal automated landing system would:

- Land any aircraft during the day or night, during any type of weather
- Land wounded or fatigued pilots
- Land in rough seas, where there was extreme carrier motion

In 1970, Bell took up the challenge and started developing such a system. Since that time, a number of different systems have been developed and installed on all U.S.A. aircraft carriers.

In addition to the carrier based systems, a number of different land based systems were also developed. One system was used to train pilots in the use of the carrier based landing system. Another system, an upgrade of the first, greatly improved reliability without sacrificing capabilities. Two of the other systems are mobile. One system can be rapidly deployed in combat conditions or it can be permanently installed at land bases. The other is small enough and rugged enough to be dropped to forward combat zones.

AN/SPN-10	First generation, aircraft landing system
AN/MPN-T1	Train personnel in operation of AN/SPN-10
AN/SPN-42A	Second generation, aircraft landing system. Uses solid state electronics, improved reliability
AN/SPN-42A-T3	Land based version of AN/SPN-42A
AN/TRN-45-MMLS	Lightweight portable ground based system
SAILS	Small lightweight system for remote sites
AN/GSN-5	U.S. Air Force version of all-weather radar landing system

As a side issue, the system nomenclature tells us exactly what type of system of it is.

AN – The following characters are part of the Joint Army-Navy Nomenclature System. The character codes have an official predefined meaning:

AN/SPN-10
S – Water (surface ship)
P – Radar
N – Navigation aid
10 – System no. 10

AN/MPN-T1
M – Ground mobile
P -- Radar
N – Navigation aid
T – Training system no. 1

AN/SPN-42A-T3
42A – System no. 42A
T3 -- Training system no. 3

AN/TRN-45-MMLS
T – Transportable (ground)
R – Radio
N – Navigation aid

SAILS – Simplified Aircraft Instrument Landing System

As illustrated in Section 2.8.2, Bell's microwave landing system actually has three modes of operation:

Automatic Mode – The radar system controls the automatic pilot, to provide accurate control to touchdown. However, the airspeed can be controlled automatically or by the pilot.

Semiautomatic Mode -- Radar information are displayed in the aircraft. The pilot manually flies the airplane, using this displayed information.

Talk Down mode – The radar operator gives instructions over the voice radio. The operator provides range, lateral position, and altitude data.

Bell's system provides a permanent record of the landing, for pilot review and training. The record includes:

(1) True Airspeed
(2) Closing Speed
(3) Sink Speed
(4) Impact Velocity
(5) Glide Slope Error
(6) Lateral Error
(7) Range
(8) Identity, Date, and Time for each aircraft

The system is designed to use a radar signal reflector, attached to the aircraft, to maximize the reflected signal. However, aircraft without the reflector can still be controlled by the radar, but there is a reduction in range and tracking accuracy.

Figure 2.8.2-a summarizes the key points of the carrier landing system. Section 2.8.2 to 2.8.8 list details on each of the seven systems.

Lightweight corner radar reflector – the only equipment required on the aircraft

References

(1) AN/SPN-10 Automatic Carrier Landing System, Bell Aerosystems Co..
(2) AN/GSN-5 All-Weather Multi-Purpose Landing System, Bell Aerosystems Co.
(3) AN/TRN-45 MMLS Mobile Microwave Landing System, Bell Aerosystems Co.
(4) Ira G. Ross Aerospace Museum, Buffalo, NY.   Cenkner Table

2.8.1-a   Two aircraft landing on an aircraft carrier using Bell's automated all-weather microwave landing system

142

2.8.2 Bell's AN/SPN-10 All-Weather Automatic Landing System

2.8.2-a Operation of AN/SPN-10 all-weather automatic landing system

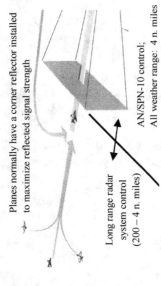

Planes normally have a corner reflector installed to maximize reflected signal strength

Long range radar system control
(200 – 4 n. miles)

AN/SPN-10 control;
All weather range: 4 n. miles

Radar window :10,000 ft wide, 700 ft high;
AN/SPN-10 locks onto airplane as it flies through

Number of planes controlled at one time -- 2
Maximum landings per hour -- 120

Touchdown is within 20 ft longitudinally and 10 ft laterally of preselected point; system compensates for ship motion

Wave-off given if system malfunctions, second aircraft is too close, or plane cannot land safely

---

Three Operating Modes of AN/SPN-10

Automatic Mode

Radar control signals are transmitted to plane autopilot to provide precision automated control to touchdown. Pilot can manually control airspeed or it can be controlled automatically, which improves performance.

Semiautomatic Mode

The pilot manually controls the plane using information displayed in the cockpit, by the radar system.

Talk-Down Mode

The AN/SPN-10 operator provides oral information to the pilot over the voice radio. The pilot lands the plane manually.

INSTALLATION ON AIRCRAFT CARRIER

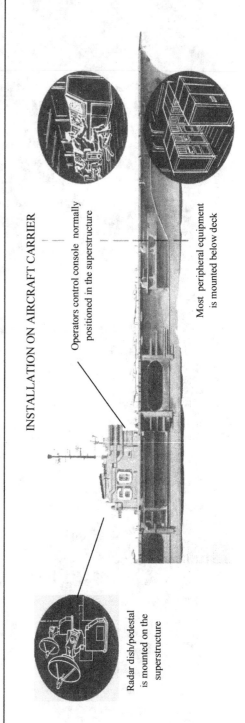

Operators control console normally positioned in the superstructure

Radar dish/pedestal is mounted on the superstructure

Most peripheral equipment is mounted below deck

2.8.2-b Bell's Data Sheet for AN/SPN-10 all weather carrier landing system

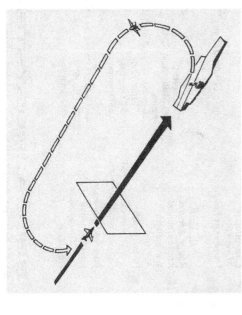

RADAR:

Range Accuracy	10 feet RMS
Angle Accuracy	1/3 milliradian RMS
Wavelength	9 mm
Antenna	48 inches – Parabolic Reflector Front End
Beam Width	
Unscanned - Max.	0.57 Degree
Scanned - Max.	0.83 Degree
Power Requirements	440 VAC, 3 Phase, 60 Cycle, 33 KW
Data stabilization:	
Angular	Pitch, Roll, Yaw
Linear	Vertical
Angular Accuracy	
Dynamic	3 Minutes of Arc RMS
Static	12 Minutes of Arc RMS
Environmental Above Deck	
Temperature	-28C to +65C
Humidity	+95% Relative

The AN/SPN-10 All-weather Carrier Landing System (ACLS) was developed by Textron's Bell Aerosystems Company for the U.S. Navy Bureau of Ships. This electronic system permits pilots to make "hands-off" landings on the pitching and rolling decks of aircraft carriers regardless of weather conditions. It is the most modern all weather landing system in existence today.

Since the first fully-automatic landing was accomplished at Niagara Falls, New York, Airport in 1954, the Bell automatic landing concept has been proven by more than 10,000 "hands-off" landings at airports, and on the decks of aircraft carriers at sea. A talk-down mode (similar to GCA) of operation incorporated in the system has been used extensively by U.S. carriers operating in Viet Nam and has increased their operating capability.

Bell's AN/SPN-10 is a closed loop system which maintains control of the aircraft from acquisition to touchdown. The major elements of the system are a precision tracking radar, data stabilization equipment, flight path computer, display and control consoles and a signal data converter.

For an automatic, "hands-off" landing, the airplane is flown to the general area of the aircraft carrier. At about four miles from the ship, precision tracking radar sets up an electronic "gate" or "window" about 10,000 feet wide, 640 feet high and 1,200 feet thick. As the airplane flies through the window, the radar locks onto the aircraft, and tracks its position as if follows a computed glide path.

The tracking radar closely observes the aircraft's flight path. When the aircraft moves off the glide slope, the radar senses this error, a computer generates a command to correct the error, and transmits it to the plane's auto-pilot which brings the aircraft back onto the glide slope.

BELL AEROSYSTEMS

A textron COMPANY

BUFFALO , NEW YORK

144

2.8.3 Bell's AN/MPN-T1 Microwave Landing System

2.8.3-a Bell's Data Sheet for AN/MPN-T1 system for operation and maintenance training of AN/SPN-10

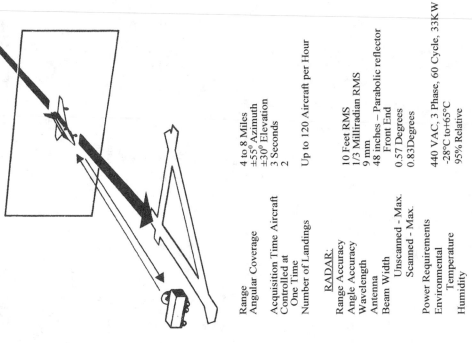

The AN/MPN-T1 Automatic Aircraft Landing System was developed by Textron's Bell Aerosystems Company for the United States Navy for training personnel in the operation and maintenance of the carrier-based, Bell AN/SPN-10 All-weather Carrier Landing System (ACLS).

Bell's AN/MPN-T1 permits pilots to make "hands-off" landings on a Mirror Carrier Landing Practice (MCLP) runway which simulates the deck of an aircraft carrier.

The AN/MPN-T1 is a closed loop system which maintains control of the aircraft from acquisition to touchdown. The major elements of the system are a precision tracking radar, flight path computer and a control console.

Since the first fully-automatic landing was accomplished at Niagara Falls, N. Y., Airport in 1954, the Bell automatic landing concept has been proven by more than 10,000 "hands-off" landings at airports, and on the decks of aircraft carriers at sea. Sixteen different types of military and civilian airplanes, including the latest jet fighters, bombers and transports, have been landed automatically.

For an automatic, "hands-off" landing the airplane is flown to the general area of the landing field. At about four miles from the field, precision tracking radar sets up an electronic "gate" or "window" about 10,000 feet wide, 640 feet high, and 1,200 feet thick. As the airplane flies through the Window, the radar locks onto the aircraft, and tracks its position as it follows a computed glide path.

The tracking radar closely observes the flight path the plane makes. When the aircraft moves off the glide slope, the radar senses this error; a computer generates a command to correct the error and transmits it to the plane's autopilot which brings the aircraft back onto the glide slope.

Range	4 to 8 Miles
Angular Coverage	±55⁰ Azimuth
	±30⁰ Elevation
Acquisition Time Aircraft Controlled at One Time	3 Seconds
	2
Number of Landings	Up to 120 Aircraft per Hour
RADAR:	
Range Accuracy	10 Feet RMS
Angle Accuracy	1/3 Milliradian RMS
Wavelength	9 mm
Antenna	48 inches – Parabolic reflector
Beam Width	Front End
Unscanned - Max.	0.57 Degrees
Scanned - Max.	0.83 Degrees
Power Requirements	440 VAC, 3 Phase, 60 Cycle, 33KW
Environmental	
Temperature	-28°C to+65°C
Humidity	95% Relative

BELL AEROSYSTEMS
BUFFALO, NEW YORK
A textron COMPANY

145

2.8.4 Bell's AN/SPN-42A Microwave Landing System

2.8.4-a Bell's Data Sheet for AN/SPN-42A all weather carrier landing system

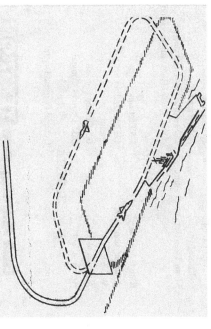

The AN/SPN-42A All-Weather Carrier Landing System (ACLS) has been developed for the U. S. Navy by Textron's Bell Aerospace Company. The ACLS provides the Navy with "zero-zero" automatic landing capability on pitching and rolling aircraft carriers. It is the most modern all-weather landing system in existence today.

The AN/SPN-42A, developed in 1966, is a digital solid-state, high reliability version of the AN/SPN-10. A digital computer and buffer unit forms the central processor and control of the AN/SPN-42A. The aircraft position data coordinate transformation, flight path generation, command control equations, stabilization for ship's motion and system management functions for two channels of simultaneous aircraft control are all performed within this complex. Precision tracking radar and stable platform sensors are mated with the computer complex to provide the precise ship's motion and aircraft position data needed for automatic control to touchdown.

The AN/SPN-42A radar scans a volume that is 25^0 wide, 1.5^0 high and 1200 feet deep, at four to eight miles aft of the carrier. The center of this volume is selected by either the shipboard Navy Tactical Data System (NTDS) or the AN/SPN-42A operator. As an aircraft enters the scanned volume, the radar acquires and tracks it, and furnishes the measured position information to the computer complex until completion of the landing.

The computer complex stabilizes the radar data for ship motion as sensed by the stable platform and accelerometer and compares the aircraft's position with a desired glide slope and center line. The computer generates flight path error signals and autopilot commands suited for the particular aircraft type under control. These commands and error signals are transmitted to control the aircraft to a completely automatic touchdown. The flight path error signals are used to drive a cross-pointer display in the cockpit and provide the pilot with a monitor of an automatic landing or the capability to fly the needles in a Mode II approach. The AN/SPN-42A operator monitors the approach and landing on his control console and has the added capability of providing a "talk-down" or carrier controlled approach as a back-up.

The AN/SPN-42A is currently deployed or scheduled for deployment on all of the U.S. Navy's modern attack carriers. Shore based versions of the equipment (the AN/SPN-42-T3 remote trainer systems) have also been deployed.

2.8.5 Bell's AN/SPN-42-T3 Microwave Landing System

2.8.5-a Bell's Data Sheet for AN/SPN-42-T3 land training system for AN/SPN-42-A

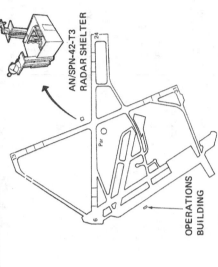

SYSTEM CAPABILITIES

Range	Touchdown to 8 Miles
Angular Coverage	± 155 Degrees Azimuth
	-5, +30 Degrees Elevation
Landing Rate	Up to 120 aircraft per hour
Simultaneous Aircraft Controlled	2
Dispersion	
Longitudinal	40 Feet rms
Lateral	10 Feet rms
Mean Time Between Failure	230 Hours (Proven, Based on Over 37,000 Hours of Operation)

RADAR

Range Accuracy	3.5 Feet rms
Angular Accuracy	1/3 Milliradian rms
Wave length	9 mm
Antenna, 4 ft	0.5 Degree Conical
Antenna, 7 ft	0.35 Degrees Conical

COMPUTER

Type	General Purpose Digital

The AN/SPN-42-T3 Automatic Landing System now gives the U.S. Navy's Air Stations a safe and positive means for landing aircraft under adverse weather conditions previously undesirable for landings.

The AN/SPN-42-T3, developed in 1970, is a digital solid-state, high-reliability, remote version of the AN/SPN-42A. A digital computer and buffer unit forms the central processor and control of the AN/SPN-42-T3. The aircraft position data coordinate transformation, flight path generation, command control equations, and system management functions for two channels of simultaneous aircraft control are all performed within this complex. Precision tracking radar provides precise aircraft position data needed for automatic control to touchdown.

The AN/SPN-42-T3 radar scans a volume that is 25° wide, 1.0° high and 1200 feet deep, at four to eight miles aft of the touchdown point. The center of this volume is selected by the AN/SPN-42-T3 operator. As an aircraft enters the scanned volume, the radar acquires and tracks it, and furnishes the measured position information to the computer complex until completion of the landing.

The computer complex compares the aircraft's position with a desired glide slope and center line. The computer generates flight path error signals and autopilot commands suited for the particular aircraft type under control. These commands and error signals are transmitted to control the aircraft to a completely automatic touchdown. The flight path error signals are used to drive a cross-pointer display in the cockpit and provide the pilot with a monitor of an automatic landing or the capability to fly the needles in a Mode II approach. The AN/SPN-42-T3 operator monitors the approach and landing on his control console and has the added capability of providing a "talkdown" or carrier controlled approach as a back-up.

The AN/SPN-42-T3 is currently deployed at Whidbey Island Naval Air Station and other shore-based earlier versions are scheduled for conversion to the AN/SPN-42-T3.

2.8.6 Bell's AN/TRN-45 MMLS Microwave Landing System

2.8.6-a   Bell's Data Sheet For AN/TRN-45 MMLS Mobile Microwave Landing System

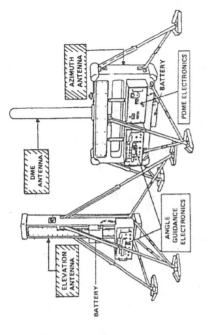

The Mobile Microwave Landing System (MMLS) is an all weather precision landing aid that can be rapidly deployed in a small, hastily established assault zone within a forward battle area.

MMLS enables re-enforcement aircraft to locate and land at temporary bases in the battle area. These bases may be jungle clearings for rotary wing aircraft or battle damaged runways for STOL transport planes.

MMLS is a mobile landing system designed by Bell Aerospace Textron in anticipation of a Department of Defense requirement, for use by all branches of the military.

- Fully interoperable with the FAA and International MLS systems for civil aircraft and airports.
- Approaches that are High Angle, Curved or Segmented are all supported by M M LS.
- Air-derived system enabling individual aircraft to select optimum approach path.
- Ground selectable coverage volume to avoid obstructions or restricted air-space.
- 200 channels, paired with Precision Distance Measuring Equipment.
- Readily transportable by land, sea and air.
- Lightweight and rugged construction for successful missions anywhere in the world.
- Deployed in less than thirty minutes by a two man team. Easily aligned and electronically leveled.
- Highly reliable system with Built-in-Test that minimizes maintenance effort.
- Microprocessor control of PDME and Angle Guidance functions.

ACCURACY		FAA CAT II
COVERAGE		
AZIMUTH		±40 Degrees
ELEVATION		+1 to +15 Degrees
RANGE		15 NM
FREQUENCY		
MMLS		X MHz
PDME		Y MHz
NUMBER OF CHANNELS		200
POLARIZATION		Vertical
SIZE		
AZIMUTH ANTENNA		36 x 50 x 6 in.
ELEVATION ANTENNA		80 x 8 x 8 in.
POWER SOURCE		Battery or external
		AC/DC power
DEPLOYMENT		Split-Site or collocated
OPERATIONAL WIND		75 Knots
MTBCF		Greater than 7000 hours
MTBCMA		Greater than 2000 hours

Bell Aerospace TEXTRON

Bell Aerospace Division of Textron Inc.

148

2.8.7 Bell's SAILS Microwave Landing System

2.8.7-a Bell's Data Sheet for Simplified Aircraft Instrument Landing System, SAILS

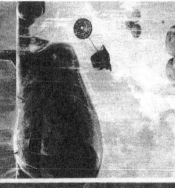

OPERATIONAL CHARACTERISTICS

Acquisition Range	
Good Weather	40 Miles
35 mm/hr Rainfall	13 Miles
Glide Slope	Selectable by pilot
Approach Heading	Selectable by pilot
Number of Aircraft	100 per Beacon
Azimuth Coverage	360^0, Ground based and airborne

GROUND STATION PARAMETERS
  Proprietary

AIRBORNE UNIT PARAMETERS
  Proprietary

SAILS, Simplified Aircraft Instrument Landing System, is a small, lightweight electronic device to aid in the landing of aircraft under low or poor visibility conditions, or at remote landing sites.

The system is comprised of lightweight ground and airborne units. Air droppable, the ground beacon is quickly and easily set-up for operation. It can be used at fixed bases or dispersed out-of-the-way locations. Maintenance is minimum. The airborne unit is mounted in a Radom, either beneath the fuselage or in the nose of an aircraft.

In operation, the airborne unit transmits a coded signal, unique to the ground unit for which it is searching. On the ground, the beacon picks up the signal and sends off a return signal on which the airborne unit locks. Aboard the aircraft, a small analog computer determines glide slope and azimuth. Standard cockpit instruments provide the pilot with ground speed, range, elevation and azimuth error signals - until visual approach can be made.

SAILS is capable of a variety of missions. In a military aspect, SAILS performs a number of functions. Not only does it enable remote area instrument landings, but it also aids in defining areas used for the low level extraction of troops and equipment. In combat SAILS serves as a homing device, guiding aircraft to their targets.

Commercial applications for SAILS can be utilized by heliports and small landing fields. This system will provide navigation and guidance for helicopters, fixed-wing and V /STO L aircraft.

Though other systems exist for guiding aircraft to remote locations, none incorporate Bell's unique "offset capability" -- its ability to vector a craft to a point some distance from the ground beacon.

149

2.8.8 Bell's AN/GSN-5 All-Weather Multi-Purpose Automatic Landing System for the Air Force

The ground based final approach landing system, AN/GSN-5, can employ three landing procedures:

(1) Ground Controlled Approach (G.C.A.) -- Standard conventional procedures that utilize manual talk down by ground control.

(2) ILS Mode – The position of the aircraft is monitored by AN/GSN-5 and the information is transmitted to the airplane. The pilot takes the required corrective action to stay on the landing path.

(3) Automatic Mode (ALS) – The position of the aircraft is monitored by AN/GSN-5, as it enters a radar window, and it is correlated with the flight path that is required for the landing airplane. This information is transmitted to the airplane autopilot and corrective action is taken automatically. To use this mode, the airplane must have an acceptable data link and an attitude-hold autopilot.

The ground controller and the pilot decide which of the three modes will be employed; the pilot can assume full control of the airplane at any time.

The system requires the use of ground equipment (Figure xxx) and airplane equipment (Figure xxx).

GROUND EQUIPMENT

Radar Trailers: Two redundant radar trailers are deployed, with each containing an independent radar tracking antenna, a beacon antenna and the required electrical equipment.

Operations Trailer: Houses the displays and controls, the flight path computer, and other equipment necessary for automatic control.

Auxiliary Equipment Trailer: UHF/VHF receivers, transmitters and portable test equipment is stored here.

Power Trailer: The system is normally controlled by a commercial power source. The power trailer supplies emergency back-up power.

AIRPLANE EQUIPMENT

Antennas: A beacon K_a-band receiving antenna, S-band transmitting antenna.

Computer: An airspeed controller and computer are necessary for automatic landing.

Standard Equipment: Autopilot, cross-pointer display, UHF or VHF communications equipment.

GENERAL SPECIFICATIONS

Range: 4 nautical miles under all weather conditions

Acquisition Gate Size at Four Nautical Miles: 10,000 feet wide x 800 feet high

Maximum Angular Coverage: $\pm 45°$ azimuth, -10 to +30 vertically

Number of Aircraft Under Control at One Time: 2

Number of Landings per Hour: 120

Overtake Protection: Automatic go-around signal will be given to overtaking aircraft.

Fail-Safe Feature: If the system becomes inoperative, the airplane under control will be given the go-around signal.

Control Volume: If the airplane is in a position from which it cannot be landed, an automatic go-around signal will be given.

Touchdown Sink Rate: 2 feet per second

Decrab   Automatic decrab is accomplished just prior to touchdown

ENVIRONMENTAL SPECIFICATIONS

Operating Temperature: -54 to +52 C

Relative Humidity: 95 percent

Wind: 60 knots operating

90 knots not operating

Primary Power Requirements: 40 kw of 120/240 volt, 60 cycle, single phase

Reference

(1) *U.S. Air Force AN/GSN-5 All-Weather Multi-Purpose Landing System*, Bell Aero-systems Company, Division of Bell Aerospace Corporation, Report no. 60001-0005R

150

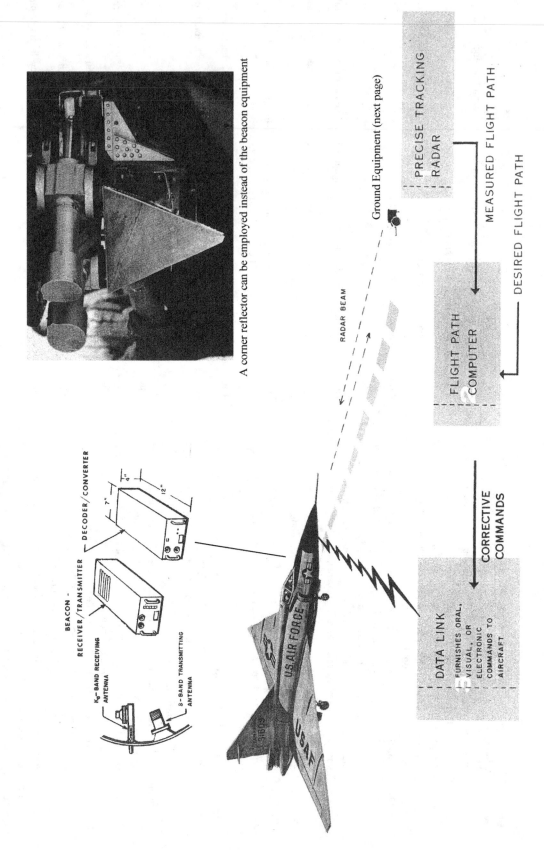

2.8.8-a   Airplane equipment required for automatic landing by the GSN-5 ALS

A corner reflector can be employed instead of the beacon equipment

Ground Equipment (next page)

PRECISE TRACKING RADAR

MEASURED FLIGHT PATH

DESIRED FLIGHT PATH

FLIGHT PATH COMPUTER

RADAR BEAM

CORRECTIVE COMMANDS

DATA LINK
FURNISHES ORAL, VISUAL, OR ELECTRONIC COMMANDS TO AIRCRAFT

BEACON -
RECEIVER/TRANSMITTER

DECODER/CONVERTER

7"
4"
12"

K$_u$-BAND RECEIVING ANTENNA

S-BAND TRANSMITTING ANTENNA

U.S.AIR FORCE

USAF

151

2.8.8-b Ground equipment for Bell's AN/GSN-5 U.S. Air Force all-weather multi-purpose landing system consists of an operations trailer, an auxiliary equipment trailer, two identical radar trailers, and a standby power trailer

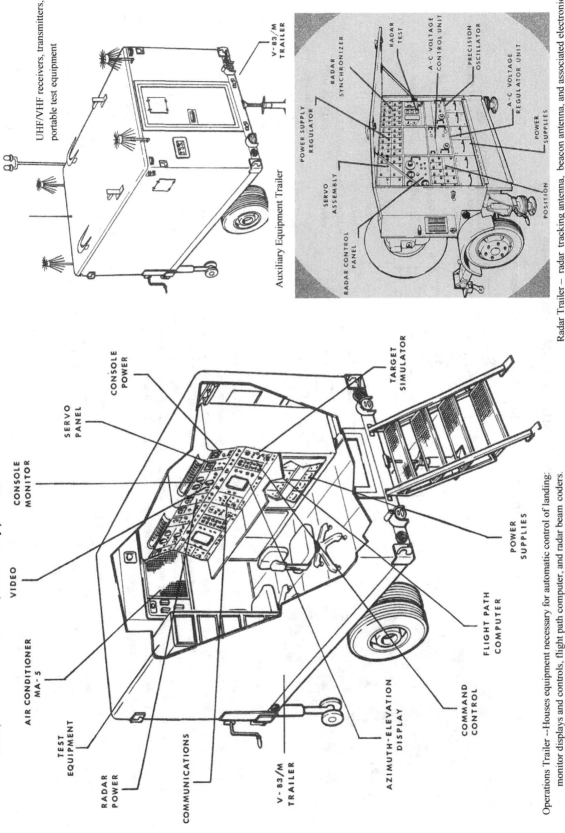

UHF/VHF receivers, transmitters, portable test equipment

V-83/M TRAILER

Auxiliary Equipment Trailer

POWER SUPPLY REGULATOR

RADAR SYNCHRONIZER

RADAR TEST

A-C VOLTAGE CONTROL UNIT

PRECISION OSCILLATOR

A-C VOLTAGE REGULATOR UNIT

POWER SUPPLIES

SERVO ASSEMBLY

RADAR CONTROL PANEL

POSITION

AIR CONDITIONER MA-5

VIDEO

CONSOLE MONITOR

SERVO PANEL

CONSOLE POWER

TEST EQUIPMENT

RADAR POWER

COMMUNICATIONS

V-83/M TRAILER

AZIMUTH-ELEVATION DISPLAY

COMMAND CONTROL

FLIGHT PATH COMPUTER

POWER SUPPLIES

TARGET SIMULATOR

Operations Trailer --Houses equipment necessary for automatic control of landing: monitor displays and controls, flight path computer, and radar beam coders.

Radar Trailer — radar tracking antenna, beacon antenna, and associated electronics

152

2.9 Airplane Communications

2.9.1 Bell's Low Profile SHF SATCOM Antenna System

2.9.1-a   Bell's Data Sheet for low profile SHF SATCOM antenna system

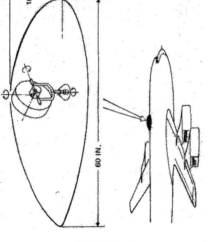

Bell Aerospace TEXTRON

SYSTEM PERFORMANCE	
Frequency	7.25 to 8.4 GHz
Gain	25 dbi
Noise Temp	75 ° K
Side lobes	13 db
Polarization	Simultaneous left and right circular
Axial Ratio	1.5 db
Isolation	20 db
Return Loss	20 db
Radiated Power level 1 kw	
Coverage	El -15° to +90°
	Az continuous 360°
Stabilization	Self contained inertial stabilization
Tracking	Conical scan
WEIGHT	
Antenna/Pedestal	65 lb
Radome	25 lb
Control Electronics	30 lb
SIZE	
Antenna/Pedestal	10 in. aperture
	15 in. swept radius
Eadome	Aerodynamic shape 16 in. high x 60 in. long
Control Electronics	½ std rack sized drawer
ENVIRONMENTAL	
Temperature/Altitude	MIL-STD-810C Method 504
EMI/RFI	MIL-STD-461A
Vibration	MIL-STD-810C Method 514.2
Explosion Proof	MIL-STD-810C Method 511.1

The Bell Low Profile SHF Antenna system is designed to meet the needs of MILSATCOM users on aircraft such as the B-52 and EC-135 where limited space is available for antenna installation. It offers a high quality circularly polarized communications antenna for duplex operation at moderate data rates with DSCS II/III. The system features automatic tracking with an inertially stabilized 2 axis pedestal providing full hemispherical coverage thereby achieving reliable communications without restricting the aircraft flight path or maneuvers. The small aerodynamically shaped radome enclosure has a minimal affect on aircraft performance and is adaptable for installation in a convenient area on top of the aircraft fuselage.

The antenna is a unique SHF Diffraction Ring device. developed by Bell specifically to achieve 26 db gain in a 10 inch aperture and form factor compatible with a small 2 axis pedestal. The resulting antenna/pedestal configuration offers a low cost, low risk approach that has the reliability and operational features required for military equipment.

2.9.2 Bell's Above-Rotor Helicopter Antenna for Tactical Satellite Communications

Bell's above-rotor super-high frequency helicopter antenna was developed to permit helicopters to become part of the Tactical Satellite Communications (TACSATCOM) network, which allows hundreds of airborne or ground units to be able to communicate via satellite.

A 10,000 mile two way voice communication was demonstrated, between an Army helicopter hovering over a New Jersey Naval Station and a control transport flying near Sydney Australia, via a tri-service tactical communications satellite. The Air Force Avionics Laboratory at Wright Patterson AF Base and the U.S. Army Satellite Communications Agency co-sponsored this contracted work.

The system utilized a 32 inch parabolic dish. It was automatically controlled by the same gyroscopes used for navigational purposes.

Traditional below-rotor helicopter antennas have suffered from a rotor induced chopping effect, during satellite communications. Bell's above-rotor antenna has solved this problem. It was determined that this configuration actually improves helicopter stability at the expense of a 6% increase in drag.

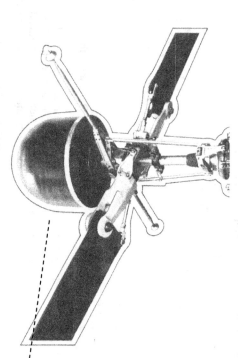

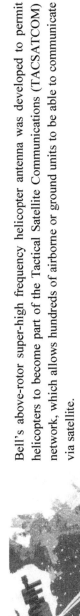

References

(1) Army Scores Communications "First" With Bell Above-Rotor Antenna, Rendezvous, Bell Aerospace, Vol. X, Spring 1971.

154

2.9.3   Bell's Visual Airborne Target Locator System (VATLS)

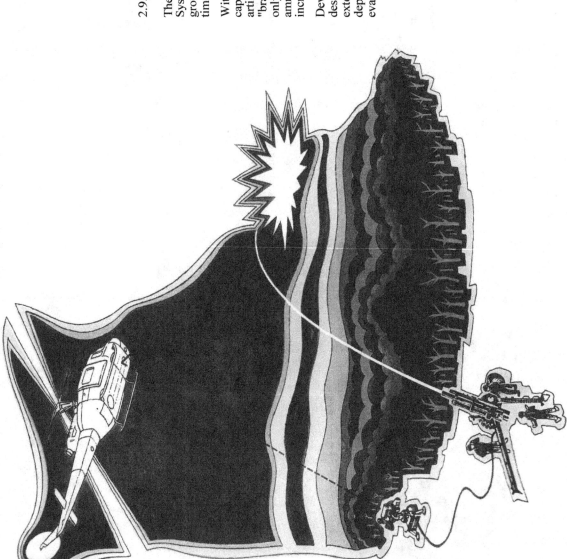

2.9.3-a   Bell's Data Sheet for U.S. Army's VATLS

The U. S. Army's Visual Airborne Target Locator System (VATLS) utilizes a helicopter and associated ground equipment to automatically pinpoint in real time enemy ground targets for artillery weapons.

With V ATLS, every round can be fired for effect, a capability that enhances the overall efficiency of artillery firepower. It's a major improvement over "bracketing" target location techniques because it not only catches the enemy by surprise, but reduces ammunition requirements and weapon wear, and increases battery security.

Developed by Bell Aerospace for the Army and designated the AN/UVS-l, VATLS underwent extensive field testing in Arizona before being deployed to Southeast Asia for a prototype test and evaluation program.

155

2.9.4 Bell's R.F. Antennas for Satellite Communication Systems

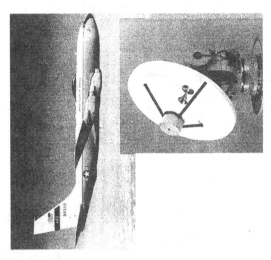

Based on experience gained with Microwave Landing Systems, Bell developed and manufactured antennas and support pedestals for three military satellite communication programs. A number of antennas were developed for use by: surface ships, submarines, land based stations, airborne command posts, and military forces. Bell's dish antennas range in size from 5 ½ to 96 inches in diameter.

(1) MILSTAR (Military Strategic and Tactical Relay) Satellite Communications System provides global military communication that is secure and jam resistant. The system utilized dual-band RF (radio frequency) antennas, operating at EHF (extremely high frequencies). Its deployment ensured reliable command, control, communications, and intelligence acquisition for strategic and tactical U.S. forces worldwide.

Bell teamed with Raytheon Equipment Division and Rockwell International in the development of the MILSTAR Terminal Program. Bell was responsible for the ground command post, airborne command post, and airborne force antenna/pedestal systems. Included were the antenna assembly, related microwave components, pedestal mechanical assembly, and control electronics.

(2) Navy's EHF Satellite Communications System. This system provides secure and anti-jam communication between surface ships, submarines, and shore stations.

(3) U.S. Army's SCOTT Program.

References

(1) Rich in Heritage, A Leader in Innovation, Bell Aerospace Textron
(2) Bell Aerospace Textron Overview, Bell Aerospace Textron

2.10 Bell's Mach 3.0 Rotating Arm Erosion Test Facility

The military identified a number of areas with severe material erosion problems that were related to atmospheric phenomenon:

(1) High speed all weather military aircraft have experienced structural damage after flying through rain. At speeds exceeding 680 mph, it occurred in thirty seconds when flying through heavy rain. Similar problems have occurred on commercial aircraft.

(2) Gas turbine airplane engines experienced erosion damage of the inlet guide vanes, in addition to the first stage of fan and compressor blades.

(3) Turbine blades of helicopter jet engines have been dramatically eroded by ingested sand, while the main rotor leading edge has been eroded while hovering, due to kicked up material.

(4) High speed missiles have also experienced these erosion problems.

(5) The reusable Space Shuttle would also be exposed to potential erosion problems.

To address these erosion problems, the Air Force Materials Laboratory contracted the Bell Aerospace Division of Textron to design and build a Mach 3.0 test facility. This facility would be used to evaluate the erosion properties of various types of aerospace materials and to develop new materials. Design work began December1966; construction was completed October 1968. This was the first rotating arm facility, worldwide, that could test at Mach 3.0. Bell had already been involved in erosion research for a number of years, having started in 1961.

The test facility was huge, measuring 26 feet in diameter, with a translating roof that could be removed on rails. Vacuum pumps could evacuate the facility to simulate pressures at an altitude of 80,000 feet. The rotating arm was mounted on a concrete slab, 20 feet below ground level. Closed circuit television was available for high speed recording of tests. A distribution system could spray rain or particles on the rotating specimen, which was mounted on the end of the rotating arm. Material specimens could be mounted at various angles, by attaching them to special holders.

References

(1) *Erosion Research at Mach 3.0*, Bell Aerospace TEXTRON

(2) *Rain and Sand Erosion up to Mach 3*, Bell Aerospace TEXTRON

(3) Anthony, F. M., Pearl, H. A., *INVESTIGATION OF FEASIBILITY OF UTILIZING AVAILABLE HEAT RESISTANT MATERIALS FOR HYPERSONIC LEADING EDGE APPLICATIONS, Vol. III, Screening Test Results and Selection of Materials* WADC Technical Report 59-744 Vol. III, July 1960, Bell Aircraft Corporation.

2.10-a Typical Test Results

Material	Threshold Velocity. ft/sec
polymers	400 Approx.
metals	700
ceramics	100 Approx.
elastomers	1100
I.R. windows	500

- Damage rates increase dramatically above threshold velocities
- Damage rates are dependent upon the type of material used; in addition to the nature, size, and velocity of impacting particles
- Erosion media include rain drops, dust, sand, and ice particles

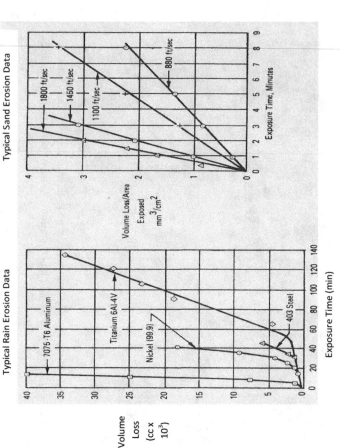

Typical Rain Erosion Data

Typical Sand Erosion Data

2.10-b   Bell's Mach 3 Rotating Arm Test Facility

90° Specimen Holder

10°/20° Specimen Holder

40° Specimen Holder

Objective

For 600 different types of aerospace materials, evaluate erosion behavior:

Speeds from Mach 0.66 to Mach 3.0
Simulated altitudes ranging to 60,000 feet

Extrapolate data to predict material life at hypersonic speeds

Test Environment

The test facility has the capability of independently controlling eight parameters that affect material erosion:

test material
exposure duration
ambient air pressure

particle material
specimen speed
angle of attack

particle size and shape
particle impingement rates

Particle Loading of Specimen

Rainfall concentration: one to three inches per hour, in an annular ring
Sand rates: one to five pounds per minute
Ice crystals: up to one pound per cubic foot per minute
Dust, bead, or other particulate: one to five pounds per minute

Vacuum chamber roof moves on rails to expose rotating arm and sample

Rotating Arm

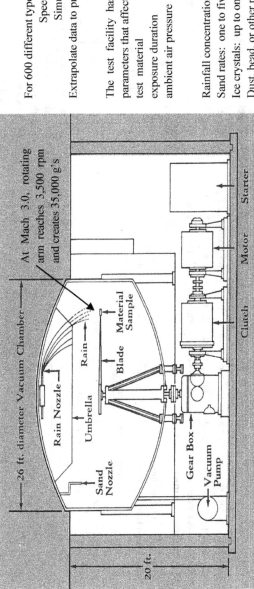

At Mach 3.0, rotating arm reaches 3,500 rpm and creates 35,000 g's

26 ft. diameter Vacuum Chamber

Rain Nozzle

Rain

Umbrella

Blade

Material Sample

Sand Nozzle

Gear Box

Vacuum Pump

Clutch   Motor   Starter

20 ft.

2.11 Air Craft Electronics

2.11.1 Inertial Navigation Systems, Accelerometers, Gravity Meters, Gyroscopes

Inertial navigation systems determine the acceleration, velocity and position of a craft using instrumentation and information stored onboard the vehicle. There is no need to access external optical or electromagnetic information, as is the case with vehicles controlled from the ground.

To develop on understanding of how a complex system operates, it's always best to begin with a simple meaningful model. We therefore consider the following system.

INERTIAL NAVIGATION SYSTEM

An inertial navigation system, that is mounted in a vehicle, is comprised of a gyroscope, an accelerometer, and a computer.

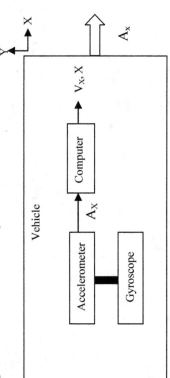

Assume that the vehicle is accelerating at the rate of "A_x" miles/hour^2, in the x direction. The accelerometer measures the acceleration, in the x direction, and transmits it to the computer. The computer then calculates the velocity and position of the vehicle. The computer also correlates the computed position with that of the desired input trajectory. If there is significant divergence, the vehicle is commanded to takes corrective action.

ACCELEROMETER

For a simplified analysis of an accelerometer, consideration has to be first given to the performance of a linear spring.

The change in length of the spring, due to the applied force, can be determined from:

$$F = K \cdot S \qquad \text{(eq. 1)}$$

Where:
F = force applied to spring
K = spring constant (determined by test)
S = change in length of spring, from no-force position

Consider next the case when a body of mass m is attached to this liner spring. From Newton's law:

$$F_{net} = m \cdot A_x \qquad \text{(eq. 2)}$$

where a force F_{net} was applied to get mass m to accelerate at A_x.

If mass m is attached to a linear spring, spring deflection S can be measured, and the spring equation can be used to compute the force applied to the spring (eq. 1). Once the force is known, Newton's law will give the desired acceleration (eq. 2) for the known mass m. Bell employs a much more sophisticated high resolution electro-mechanical accelerometer, but the basic concept is the same.

GYROSCOPE

The gyroscope is a sophisticated device that keeps the accelerometer oriented in the X direction, regardless of how the vehicle changes orientation.

For a vehicle moving in three-dimensional space, three independent gyroscopes and accelerometers are required. Each gyroscope maintains its accelerometer mass pointed in one of the X, Y, or Z directions, so the velocity component in that direction can be measured, as can the corresponding vehicle position.

GRAVITY- METERS

The accelerometer can also be used to measure the local magnitude and direction of gravity, using three accelerometers/gyroscopes. If the vehicle is moving in the X direction at a uniform speed, $A_x = 0$; Newton's law gives $F_{net} = 0$.

The only force acting on the accelerometer mass m is therefore gravity, $F_X = mg_X$. If the vehicle is moving through a gravity field at constant velocity, the local magnitude and direction of the field can be determined from the above relationships, and the entire gravity field can be mapped.

References

(1) *EXPLORING IN AEROSPACE ROCKETRY*, U.S. Government Printing Office, 1971.
(2) Considine, D. M., *ENCYCLOPEDIA OF INSTRUMENTATION AND CONTROL*, NcGraw-Hill Book Company, 1971.

2.11.1-a   Inertial Navigation Systems, Accelerometers, Gravity Meters, Gyroscopes

Function	Name	Comment
High Performance Navigation System	HIPERNAS	3 BRIG Gyros and 3 Model III-B accelerometers
Miniature Electrostatic Accelerometer	MESA	Single axis linear accelerometer
Bell Miniature Precision Accelerometer		Models III-B, IV, V, VI, VII, IX, XI
Bell Rotating Inertial Gyro	BRIG	Two degree of freedom gyroscope

2.11.1-b   Gravity Sensing Systems. Gravity measurement can provide information for navigation, weapons delivery and resource exploration

Function	Name	Comment
Bell Gravity Meter	BGM-1, BGM-2	Proved that inertial grade accelerometers could significantly improve the shipboard measurement of gravity. Upgraded to BMG-3
Bell Gravity Meter	BGM-3	Dynamic gravity measurement. Used for magnitude of gravity mapping surveys.
Gravity Sensor System	GSS	Navigation of U.S. Navy submarine launched Trident missile and submarines.
Gravity Gradiometer Survey System	GGSS	Gravity survey from aircraft or land

Cenkner Tables

2.11.2 Bell's High Performance Navigation System (HIPERNAS)

HIPERNAS is a sophisticated inertial navigation system that was developed by Bell Aerosystems Company.

HIPERNAS II employs three Brig II gyroscopes and three Model IIIB accelerometers, to provide three dimensional navigation information. It was employed by the U.S. Air Force for navigation and camera stabilization during mapping and survey missions.

HIPERNAS can also be used to guide a missile to a predetermined target, to guide spacecraft to any point in the Universe, or to control a ship or submarine on a known course to any water destination.

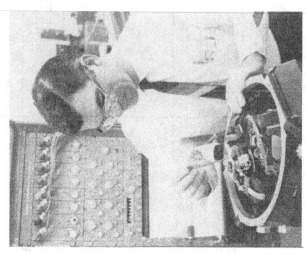

2.11.2-d Assembly of HIPERNAS

2.11.2-a HIPERNAS
Platform

2.11.2-c Brig II Gyroscope

2.11.2-b HIPERNAS Electronics Support

2.11.3 Bell's Miniature Electrostatic Accelerometer (MESA)

2.11.3-a Bell's Data Sheet for Miniature Electrostatic Accelerometer

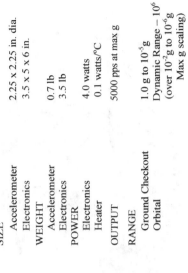

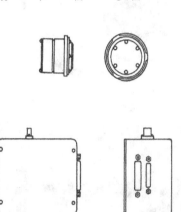

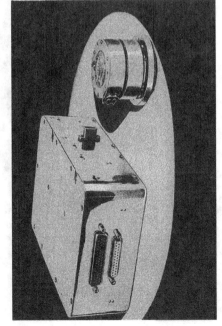

SPECIFICATION		
SIZE		
	Accelerometer	2.25 x 2.25 in. dia.
	Electronics	3.5 x 5 x 6 in.
WEIGHT		
	Accelerometer	0.7 lb
	Electronics	3.5 lb
POWER		
	Electronics	4.0 watts
	Heater	0.1 watts/°C
OUTPUT		5000 pps at max g
RANGE		
	Ground Checkout	1.0 g to 10^{-5} g
	Orbital	Dynamic Range – 10^6 (over 10^{-2} g to 10^{-6} g Max g scaling)

PERFORMANCE		
BIAS		
	Nominal	0.01% of max g
	Repeatability	0.001% of max g
	Stability	0.001% of max g
SCALE FACTOR		
	Stability	0.1%
	Linearity	0.1%

ENVIRONMENT (NON-OPERATING)	
TEMPERATURE	-65°F to 160°F
SHOCK	40 g for 11 msec (typical)
VIBRATION	7.5 g peak 20-2000 cps (typical sine)
	0.07 g^2/cps 20-2000 cps (typical random)

The Bell Miniature Electrostatic Accelerometer (MESA) is a single axis linear accelerometer designed to meet new systems requirements for space age vehicles. MESA offers the possibility to measure accelerations which are one millionth of the maximum acceleration experienced in a specified space environment.

The instrument utilizes the principle of electrostatic suspension and constraint on all three axes. These capabilities can be raised or lowered by adjusting voltages in the support electronics. Measurement of acceleration along the single axis is achieved by a small two-inch diameter transducer which is qualified for space environments.

Designed as a rugged instrument for launch environment, the key elements are built of beryllium in a simple geometric configuration. The one moving part is a cylinder which is supported from within by eight capacitive support pads and is gas damped in a hermetically sealed aluminum housing. The accelerometer electronics are derived from the Bell Aerosystems line of reliable digital velocity meters which are used for the Agena, Scout, Minuteman-Mark 11 RV and many other space programs.

Bell's MESA can be launched in a non-operating mode. Operation in flight is initiated by a single command which activates resonant circuit electrostatic suspension and pulse rebalance acceleration measuring circuitry. Automatic recycling is provided in the event of temporary acceleration overload of the system. Automatic range changing is available.

The MESA provides a check-out range to allow calibration of the instrument in the earth's gravity field and a low "g" range for operation in an orbital environment. The military and NASA missions considered are orbital drag measurements, ion jet thrust measurement, attitude control, direct measurement of planetary gravity gradients, and the measurement of spacecraft disturbances.

BELL AEROSYSTEMS

BUFFALO, NEW YORK · A [Textron] COMPANY

162

2.11.4 Bell's Miniature Accelerometers

2.11.4-a Various types of Bell miniature precision accelerometers for inertial guidance and navigation systems

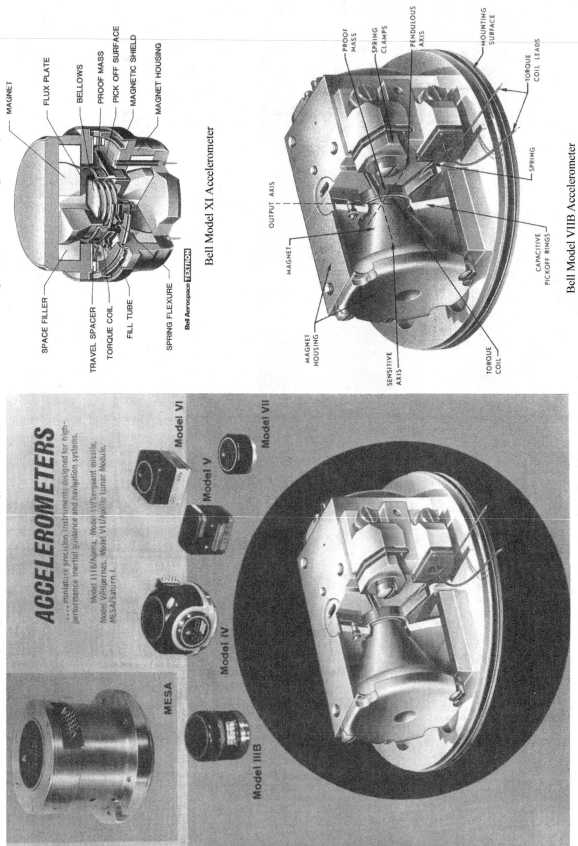

Bell Model XI Accelerometer

Bell Aerospace TEXTRON

Bell Model VIIB Accelerometer

163

2.11.5 Bell's Rotating Inertial Gyro (BRIG) Model IIB

2.11.5-a Bell Data Sheet for Rotating Inertial Gyro Model IIB

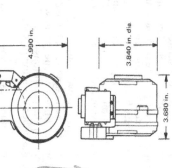

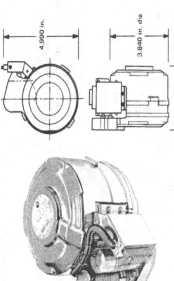

The Bell Rotating Inertial Gyro (BRIG) Model IIB is a high-performance, floated, two-degree-of-freedom gyro using the case rotation principle. Over 13 years of design, development and manufacturing experience have resulted in this small, light-weight, reliable, high performance gyro.

The major components of the B RIG II Bare:

(1) a self-acting, hydrodynamic; gas lubricated rotor bearing (as part of the hysteresis synchronous spin motor)
(2) a two-axis inductive pickoff
(3) a two-axis electromagnetic torquing system
(4) a rotating case system

The torquing system consists of a set of air-core torquing coils reacting against the field produced by a set of permanent magnets on the float. Temperature control provisions include six thermistors, resistance float sensors, two maintenance (control) heaters and a warm up heater.

The rotating case system virtually eliminates torques produced by the suspension system, signal generator, torque reaction and radial mass unbalance. Case rotation is accomplished with a hysteresis synchronous motor and a slipring/brush block assembly. Relationship between the rotating pickoff signals and the fixed case torques is provided by a resolver geared to the rotating case. The pickoff circuitry package is mounted to the brush block. Rotating and fixed magnetic shields are provided to reduce sensitivity to external magnetic fields.

The excellent drift rate performance and repeatability of the BRIG make it ideally suited for inertial navigation, attitude reference, and precision gyro compassing systems. Programs using the BRIG-series gyros are:

- Hipernas II (Air Force)
- AN/USQ-28 Geodetic and Mapping System
- AN/APQ-108 Motion Measurement and Navigation System
- AN/UVS-1 Visual Airborne Target Locator System
- Skyscraper System (MIT, Lincoln Labs)
- Project Defender (AVCO)
- Gyroscopic Azimuth Aligner (Boeing)
- Hipernas III/SARCALM for ADO-51

GENERAL

Weight (lb)	4.6
Angular Momentum (gm-cm /sec^2)	$1.3 \, (10)^6$
Output Axis Freedom (sec)	± 1000
Operating Temperature (C)	80

ENVIRONMENTAL

Vibration MIL-E-5400M, *Curve* III	
Shock (g for msec duration)	15 at 11
Angular Acceleration (rad/sec^2)	25
Operating Life (hours)	20,000

PICKOFF

Pickoff Scale Factor (mv/sec)	6.0
Excitation (volts-rms)	30.0
Frequency (Hz)	24,000

TORQUER

Scale Factor (°/hr/ma)	2.5
Linearity (%)	± 0.01

MOTOR

Synchronous Speed (rpm)	24,000
Excitation Frequency - 3 phase (Hz)	1500
Power, Running (watts)	11.0

HEATER

Warm-up Heater Power (watts)	75
Maintenance Heater Power (watts)	50 max
Temperature Sensor (ohms)	310

DRIFT

Acceleration-Insensitive Drift, Long Term, Random (output axis vertical) (10) (°/hr)	0.00019
Day-to-day Stability (°/hr)	0.00023
Acceleration-Sensitive Drift, Long Term, Random (input axis vertical) (10) (°/hr)	0.0013
Day-to-day Repeatability (°/hr/g)	0.025
Anisoelasticity (°/hr/g^2)	0.05
Stabilization time from room temperature (minutes, max.)	40, S.A. Horiz / 20, S.A. Vert

2.11.6   Bell's Model IX Accelerometer

2.11.6-a   Bell's Data Sheet for Model IX Accelerometer

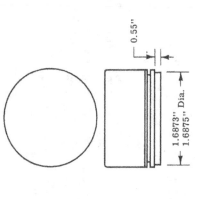

SIZE	1-3/4 in. dia. x 1 in. high (Transducer with Rebalance Electronics)
WEIGHT	Less than 5 oz.
POWER	±15V DC, less than 1 watt
RANGE	± 1 x 10^{-4} Full Scale to ± 30 g Full Scale
STABILITIES	
Null	Better than 1 x 10^{-4} g
Scale Factor	Better than 0.005%
Linearity	Better than 0.01% of g's measured
ENVIRONMENT	
Vibration	Greater than 26 g RMS
Shock	200 g's

Bell's MODEL IX ACCELEROMETER is small, low-cost, and lightweight, single-axis pendulous-type force balance accelerometer with the sensor and force balance-electronics (analog), all in one package. With the addition of a pulse-rate converter or pulse-rebalance section the MODEL IX ACCELEROMETER becomes relevant for digital operations as well. This accelerometer is particularly suitable for use in low-cost missiles, navigation systems (gimbaled or strap down), space missions, instrumentation and shipboard applications.

The heart of the Bell MODEL IX ACCELEROMETER is the pendulous proof mass consisting of a circular metallic coilform, to which two supports are attached, and a forcer coil wound on the coilform. The proof mass is suspended on two thin springs in a high intensity radial magnetic field produced by a permanent magnet contained in a soft iron armature. The circular flux gap is established by the armature and circular soft iron plate attached to the magnet. The form carrying the forcer coil acts as the center plate of a three plate variable capacitor. Two immobile insulated metal rings, one of which is placed on each side of the flux gap, form the outer plates of the three plate capacitor. The latter forms part of a capacitance bridge which is excited with alternating current. When under the influence of acceleration and/or gravity, the proof mass moves out of its center position, the capacitance bridge is unbalanced. The resulting AC signal is fed into an amplifier demodulator into a DC amplifier and returned to the forcer coil as a DC current. This current, which generated a force which constrains the proof mass towards its center position, is proportional to the acceleration in equilibrium. The output signal of the system is represented by the voltage created across a precision resistor placed in series with the forcer coil.

The required system damping is supplied through two sources: the eddy current flowing through the torque coil frame (acting like a shorted turn in the magnetic field) and, the derivative feedback provided by the loop closing amplifier. In open loop condition (whenever power is not applied) only eddy currents contribute to the damping of the movements of the pendulous mass. This damping is high and effectively protects the pendulous mass against damage under vibration or shock.

Special design considerations include such features as, a closed magnetic path around the pendulum to minimize flux leakage and to protect the transducer from external fields and allowing the mounting of several transducers in close proximity. The complete loop-closing electronics are incorporated within the accelerometer housing to provide a compact one piece sensing system.

0.55"

1.6873"
1.6875" Dia.

2.11.7 Bell's Gravity Sensor System (GSS)

2.11.7-a   Bell's Gravity Sensor System, developed for the U.S. Navy, proved to be a major improvement in navigational accuracy. It became an integral part of the Trident II submarine navigational subsystem, by correlation of measured local gravity with gravity maps.

FBM AND ATTACK SUBMARINES

**BINNACLE AND
ELECTRONICS**

STABILIZED PLATFORM

Bell Aerospace TEXTRON

2.11.8 Bell's Gravity Gradiometer Survey System (GGSS)

2.11.8-a Bell developed, built, and tested the Gravity Gradiometer Survey System (GGSS) – a variant of the U.S. Navy GSS – for the Defense Mapping Agency. It provides a method for high speed, highly accurate, gravity mapping using airborne or mobile land vehicles. Flying over a survey area, at a 2000 foot altitude, 2000 linear kilometers of gravity data can be acquired, per day. The data would be available for navigational purposes and for the identification of the location of resources.

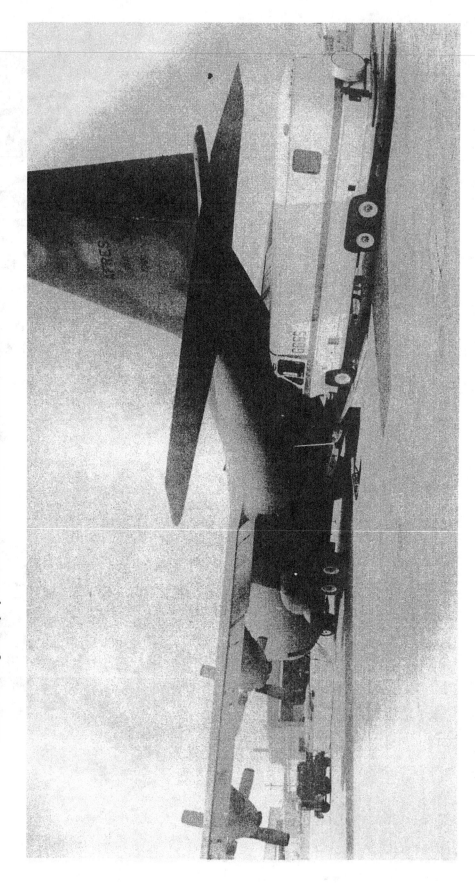

167

2.11.9 Stabilized Optical Tracking Device (SOTD)

2.11.9-a Bell's Data Sheet for SOTD

SOTD stands for Stabilized Optical Tracking Device. It also means that co-pilot gunners of high-speed, highly-maneuverable Army helicopters can locate, pinpoint and train their sights on enemy targets with speed and precision. SOTD is a universal, multi-purpose sight that functions as a gun, missile or observation sight that integrates with fire control systems for any weapon. It allows use of forward firing weapons or moving turrets mounted in front. SOTD permits unrestricted co-pilot flying of the helicopter.

Now ready for production for the U.S. Army, this versatile tracking device is inertially stabilized and virtually isolated from vibration, permitting the use of high magnification optics to aid in the tracking of distant and moving targets. Narrow and wide fields of view arc provided for enemy surveillance. The unit is roof-mounted on the co-pilot side. The eyepiece is stowed when not in use.

SOTD is an outgrowth of Bell's extensive experience in the field of inertial guidance. Other stabilized sighting devices for command guidance of helicopter launched missiles and a highly sophisticated system for a Visual Airborne Target Locating System (VATLS) still are either active programs representative of Bell's broad inertial guidance capability,

Reference

(1) SOTD stands for Stabilized Optical Tracking Device, Rendezvous, Bell Aerosystems. Vol. IV. No. 6. 1965

2.11.10 Communication Interference Problems

With so many electronic devices deployed on the battle field the army encountered, during WW II, electronic interference between these devices and they experienced a severe reduction in effective communications.

The U. S. Army established the Electronic Proving Ground, at Fort Huachuca, with a mission to solve this interference problem and to make sure that all electronic equipment performed the way it was supposed to, on the battlefield.

Included at the Fort Huachuca complex were two major test areas: the Electro-magnetic Environmental Test Facility (EMETE) and the System Test Facility (STF). Bell was contracted to operate, maintain, and develop both of these facilities.

Contractually, Bell was obligated to identify the source of this interference and to develop methods to eliminate it.

To meet their objectives, Bell developed a sophisticated computer program – the Interference Prediction Model (IPM) – that could be used to simulate any type of military equipment being used in a battlefield, anyplace in the world. Simulation included terrain, weather, foliage, transmitter and receiver data, in addition to enemy electronic equipment and jamming equipment.

The EMETE field test site is used to: acquire data for the IPM, evaluate new equipment in operational simulations, and perform special field tests.

In addition to working on eliminating communication interference problems, the STF was involved in acceptance evaluation of: surveillance, navigation, electronic guidance, and control equipment.

2.11.10-a Bell Control Center at Fort Huachuco

Reference

(1) Unscrambling the Army's Communications Interference Problem,Rendezvous, Bell Aerosystems, Vol. IV, No. 6, 1965.

2.12 Computer Software Systems

2.1.12-a   Bell's computer software systems. In the 1970's Bell began working with hybrid analog/digital computers, minicomputers, and microprocessors utilizing numerous computer languages: Ada, Pascal, FORTRAN, and Assembly. To this point, it included DEC, DG, HP, 68000, INTEL, TI, IBM + MIL-Spec computers.

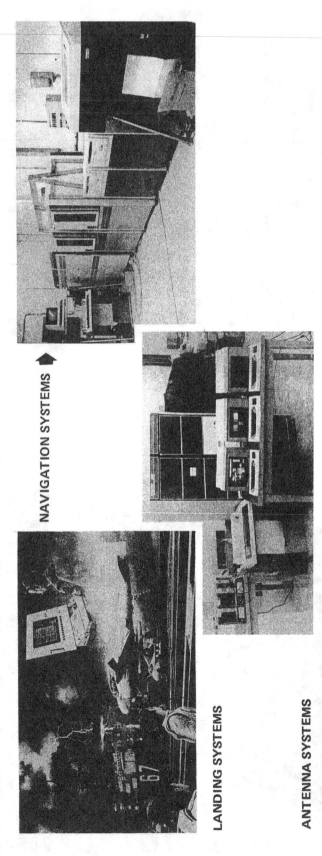

NAVIGATION SYSTEMS

LANDING SYSTEMS

PROCESS CONTROL & GRAPHICS SYSTEMS

SURVEY SYSTEMS

ANTENNA SYSTEMS

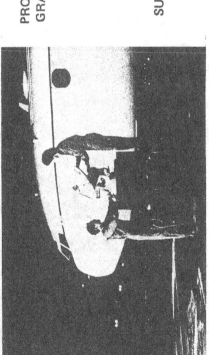

169

3.0 Space Craft

3.1 Bell's Storage Tanks Used on the Apollo Lunar Lander Spacecraft

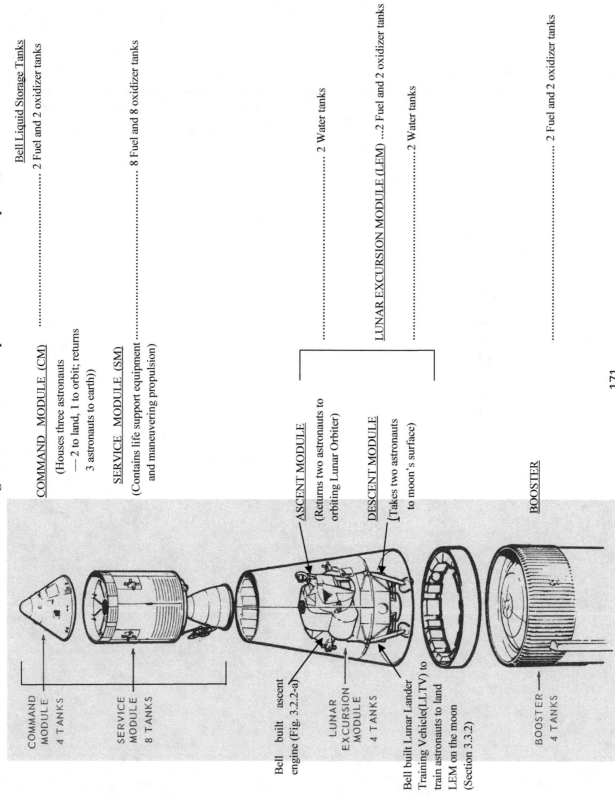

Bell Liquid Storage Tanks

COMMAND MODULE (CM) .... 2 Fuel and 2 oxidizer tanks

(Houses three astronauts
— 2 to land, 1 to orbit; returns
3 astronauts to earth))

SERVICE MODULE (SM) .... 8 Fuel and 8 oxidizer tanks
(Contains life support equipment
and maneuvering propulsion)

ASCENT MODULE .... 2 Water tanks
(Returns two astronauts to
orbiting Lunar Orbiter)

DESCENT MODULE .... 2 Water tanks
(Takes two astronauts
to moon's surface)

LUNAR EXCURSION MODULE (LEM) ...2 Fuel and 2 oxidizer tanks

BOOSTER .... 2 Fuel and 2 oxidizer tanks

COMMAND
MODULE
4 TANKS

SERVICE
MODULE
8 TANKS

Bell built ascent
engine (Fig. 3.2.2-a)

LUNAR
EXCURSION
MODULE
4 TANKS

Bell built Lunar Lander
Training Vehicle(LLTV) to
train astronauts to land
LEM on the moon
(Section 3.3.2)

BOOSTER
4 TANKS

171

3.1.1 Operation of Bell's Positive Expulsion Tanks

3.1.1-a Bell's Data Sheet for Positive Expulsion Bellows Tanks

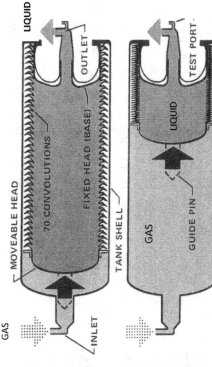

Under zero gravity or less than earth-gravity conditions of space, rocket propellants and other fluids become independent. Inside a partially-filled container, liquids may float around in clumps, bunch up at one end or cling to the walls. The method developed to overcome this critical deficiency is called a positive expulsion, which forces out and assures maximum utilization of all liquid in the tank.

Textron's Bell Aerosystem has developed, and is producing, a unique positive expulsion device, called the "positive expulsion bellows." Its capabilities include center-of-gravity control, zero permeation, simplified liquid-level sensing, storability, reproducible mode of expulsion, slosh-damping and dynamic stabilization.

This version has the stainless-steel outer shell enclosing the bellows, containing fuel or oxidizer, which collapses under the pressure exerted by externally introduced, compressed inert gas, between the shell and the bellows. Under this pressure, the moveable head travels axially toward the outlet at the fixed end of the bellows, providing a positive bubble-free supply of propellant to the rocket-thrusters, on demand. The shell's surface is smooth, allowing the bellows to slide without hindrance. Sensors on the outer shell indicate the quantity of fuel or oxidizers in the tanks. Because the bellows remains under pressure, propellant-flow to the rocket thrusters is immediately answered by a demand-valve positioned at the outlet orifice.

The positive expulsion bellows have consistently delivered 99 percent of the liquid from the tanks on spacecraft, including the Gemini-Agena Target Vehicle. Cycle-life of 200 cycles has been demonstrated.

A typical positive expulsion tank ... designed, built and flight qualified by Textron's Bell Aerosystems Company for various space programs ... consists of a stainless steel shell and a flexible steel bellows. The bellows contains the liquid.

When the liquid -- be it fuel or oxidizer -- is required, compressed helium or nitrogen gas (dotted arrow) is forced against the bellows moveable head, pushing the bellows and its contents (solid arrow). The propellant, in turn, is forced through the outlet to the rocket thrusters.

The flow of propellant can be shut off at any time by closing a valve between the tank outlet and the rocket thruster. Because the bellows remains under pressure, propellant flow can be resumed instantly by reopening the valve.

3.1.1-b Bell's Bladder Type Positive Expulsion Tank

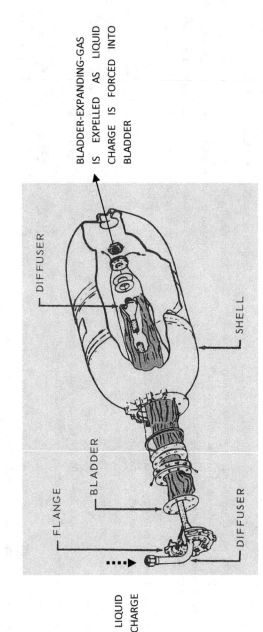

FLANGE

BLADDER

DIFFUSER

DIFFUSER

SHELL

LIQUID
CHARGE

BLADDER-EXPANDING-GAS
IS EXPELLED AS LIQUID
CHARGE IS FORCED INTO
BLADDER

These bladder type storage tanks are cylindrical or spherical in shape, with the cylindrical tanks having rounded ends. Inside each tank shell is a flexible Teflon bladder that holds the stored liquid (propellant or water). A 3/8 inch aluminum tube, the "diffuser", runs down through the center of the bladder. The diffuser is perforated with numerous drilled holes that are 0.032 and 0.040 inches in diameter. For fuel tanks, 746 of these holes are positioned in a pre-determined pattern. For oxidizer tanks, this is increased to 830 holes.

To fill the tank with a liquid, gas is first fed into the bladder through the diffuser -- to expand the bladder. The liquid charge is then introduced through the same diffuser, while the bladder-expanding-gas is bled out through a port at the other end.

To expel the liquid from the bladder, high pressure nitrogen is fed into the gap between the shell and the bladder. This forces the liquid back out through the diffuser.

If the nitrogen pressure is maintained on the outside of the bladder, a valve on the diffuser can be turned on and off to provide instantaneous control of the liquid flow.

This is especially important if the tank is supplying propellant to a rocket engine. The engine can be started or stopped by simply opening or closing the valve

The Lunar Orbiter – which orbits the moon at a altitude of 28 miles – used two cylindrical type bladder tanks for 9 gallons of fuel each and two to hold 11 gallons of oxidizer each. The orbiter will photograph the moon's surface to identify possible future landing sites, while the other two astronauts are on the surface. The tanks are approximately a foot in diameter, with the fuel tanks being 17 inches long and the oxidizer tanks 20 inches long. All tanks weigh between 7 and 8 pounds, when empty. The shell material is strong, but light, titanium .

To fabricate the bladder, a metal form is made with the same outside diameter as the inside diameter as the bladder. Atomized liquid Teflon is sprayed on the form as it rapidly rotates. Multiple heat curing and spray steps are required to build up the proper bladder thickness.

Reference

(1) *How Positive Expulsion Tanks Work*, Rendezvous, Bell Aerosystems, Vol. V / No.3 / 1966.

3.1.2　Bell's Apollo Lunar Module Propellant Tanks

3.1.2-a　Bell's Data Sheet for Apollo Lunar Module propellant tanks

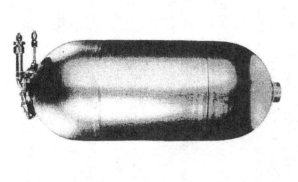

The positive expulsion propellant tanks for the reaction control subsystem of Project Apollo's Lunar Module (LM), are designed, built and tested by Textron's Bell Aerosystems Company. Bell, the foremost builder of positive expulsion devices in the aerospace industry, was selected to develop the LM tankage by Grumman Aircraft Engineering Corporation, NASA's prime contractor for the LM spacecraft. The LM is the portion of the three-module Apollo spacecraft which will land on the moon.

Bell tanks will be utilized in the LM spacecraft to supply the propellants to the reaction control rockets. These rocket thrust chambers are used for positioning, orientation and stabilization of the LM during its descent flight to the lunar surface and ascent flight to the lunar orbiting Apollo spacecraft. The LM reaction control subsystem propellant tankage consists of two fuel and two oxidizer tanks. Each tank is a cylindrical metal shell containing a flexible Teflon bladder and a perforated diffuser assembly, for positive expulsion of the propellants. A pressurization port is provided on the tank shell and a bladder bleed port is incorporated in the diffuser assembly.

Positive expulsion systems are a necessity for space vehicles, because liquid propellants do not flow naturally toward a tank outlet as they would on earth. Instead, under the zero, or less than earth gravity conditions of space flight, the propellants tend to float in the tank, or cling to the tank walls. When propellant is required pressurized helium gas is forced into the tank assembly between the tank shell and bladder. The pressurizing gas displaces the propellant by collapsing the bladder about the central perforated diffuser tube, thereby forcing the propellant through the tube, and out the discharge port of the tank. This system ensures delivery of propellants to the propulsion system in all attitudes and gravitational fields.

BELL AEROSYSTEMS

A [textron] COMPANY

BUFFALO, NEW YORK

174

3.1.3 Bell's Propellant Tanks for the Saturn S-IVB Launch Vehicle

3.1.3-a Bell's Data Sheet for Propellant Tanks Used on Apollo Spacecraft Saturn S-IVB Launch Vehicle

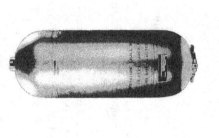

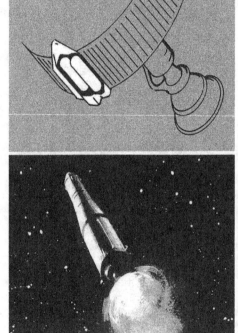

Textron's Bell Aerosystems Company, the foremost developer of positive expulsion systems was selected to provide the positive expulsion propellant tanks for the auxiliary propulsion system of the Saturn S-IVB vehicle by Douglas Aircraft Company's Missile and Space Systems Division, principal contractor on the S-IVB vehicle for NASA's Marshall Space Flight Center. The Saturn S-IVB is the third stage of the three stage Saturn V launch vehicle which will propel the manned Apollo spacecraft to the moon.

The Bell positive expulsion tanks will supply propellants to the attitude control engines during powered flight, earth orbit, and translunar coast.

Positive expulsion systems are a necessity in space vehicles because liquid propellants do not flow naturally toward a tank o utlet as they would on earth. Instead, under the zero, or less than earth gravity conditions of space flight, the propellants tend to float in the tank, or cling to the tank walls.

The S-IVB auxiliary propulsion system propellant tankage consists of two fuel and two oxidizer tanks. Each tank is a cylindrical me tal shell containing a flexible Teflon bladder and a perforated diffuser assembly for positive expulsion of the propellants. A pressurization port is provided on the tank shell and a bladder bleed port is incorporated in the diffuser assembly. When propellant is required, helium gas is forced into the tank assembly between the tank shell and bladder. The pressurizing gas displaces the propellant by collapsing the bladder about the central perforated diffuser tube, thereby forcing the propellant through the tube and out the discharge port of the tank. This system ensures delivery of propellants to the propulsion system in all attitudes and gravitational fields.

BELL AEROSYSTEMS

BUFFALO, NEW YORK   A textron COMPANY

175

3.1.4 Bell's Propellant Tanks Used on Apollo Command and Service Modules

3.1.4-a   Bell's Data Sheet for reaction control tanks used on Apollo Command & Service Modules

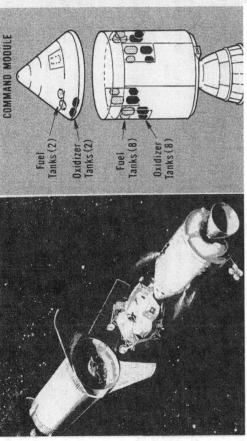

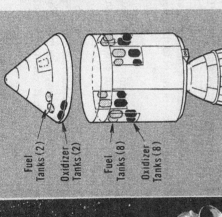

COMMAND MODULE

Fuel Tanks (2)
Oxidizer Tanks (2)
Fuel Tanks (8)
Oxidizer Tanks (8)

(SM) OX   (SM) FUEL   (CM) OX   (CM) FUEL

Textron's Bell Aerosystems Company, the foremost developer of positive expulsion systems, is providing the positive expulsion tankage for the Apollo spacecraft's reaction control system. Bell was selected to develop the tankage by North American Aviation's Space Division, principa l contractor on the Apollo Command and Service Modules for NASA's Manned Spacecraft Center.

Bell tanks are utilized in the Command and Service Modules to supply the propellants to the reaction control rockets. These rocket thrust chambers are used for positioning, orientation, and stabilization of the spacecraft during the flight to and from the Moon, and during the reentry maneuvers.

The Command Module (CM) reaction control subsystem propellant tankage consists of two fuel and two oxidizer tanks; whereas the Service Module (SM) contains 8 fuel and 8 oxidizer tanks. Four fuel and four oxidizer tanks in the Service Module are of the Command Module configura - tion.

Each tank is a cylindrical metal shell containing a flexible Teflon bladder and a perforate d diffuser assembly for positive expulsion of the propellant. A pressurization port is provided on the tank shell and a bladder bleed port is incorporated in the diffuser assembly.

When propellant is required, helium gas is forced into the tank assembly b etween the tank shell and bladder. The pressurizing gas displaces the propellant by collapsing the bladder about the central perforated diffuser tube, thereby forcing the propellant through the tube and out the discharge port of the tank. This system ensures delivery of propellants to the propulsion system in all attitudes and gravitational fields.

FUEL TANKS
PROPELLANT: Monomethylhydrazine

DIMENSIONS	
SM Diameter:	12.5 inches
SM Length:	23.7 inches
CM Diameter:	12.5 inches
CM Length:	17.3 inches
WALL THICKNESS:	
SM	0.022 inch
CM	0.027 inch
OPERATING PRESSURE:	
SM	179 psia
CM	289 psia
WEIGHT:	
SM	7.9 lbs
CM	7.2 lbs

OXIDIZER TANKS
PROPELLANT: Nitrogen Tetroxide

DIMENSIONS	
SM Diameter:	12.5 inches
SM Length:	28.5 inches
CM Diameter:	12.5 inches
CM Length:	20.0 inches
WALL THICKNESS:	
SM	0.022 inch
CM	0.027 inch
OPERATING PRESSURE:	
SM	179 psia
CM	289 psia
WEIGHT:	
SM	8.7 pounds
CM	7.9 pounds

All tanks have Teflon bladders, 9-20 expulsion cycles demonstrated; 91.5%-99% expulsion efficiency demonstrated.

BELL AEROSYSTEMS
BUFFALO , NEW YORK   A textron COMPANY

3.1.5 Bell's Water Tanks for Apollo Lunar Module

3.1.5-a Bell's Data Sheet for Apollo Lunar Module water tanks

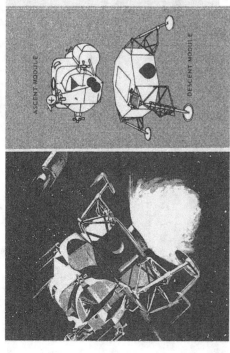

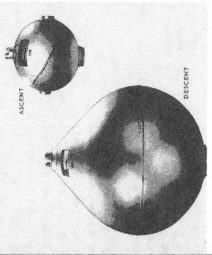

SPECIFICATIONS	
ASCENT WATER TANK	
PRESSURANT	Nitrogen
DIMENSIONS	
Diameter	14.7 inches
OPERATING PRESSURE	54 psia
WALL THICKNESS	0.030 inches
WEIGHT	6.0 pounds
BLADDER MATERIAL	Silicone Rubber
CYCLE LIFE	50 Expulsions
EXPULSION EFFICIENCY	99%
DESCENT WATER TANK	
PRESSURANT	Nitrogen
DIMENSIONS	
Length	32.5 inches
Diameter of Sphere	28.6 inches
OPERATING PRESSURE	54 psia
WALL THICKNESS	0.041 inch
WEIGHT	22.8 pounds
BLADDER MATERIAL	Silicone Rubber
CYCLE LIFE	50 Expulsions
EXPULSION EFFICIENCY	99%

11-7202-2

Textron's Bell Aerosystems Company was selected by Hamilton Standard Division of United Aircraft Corporation to develop the positive expulsion water tanks for the environmental control system of Project Apollo's Lunar Module (LM). The LM is that portion of NASA's three-module, manned spacecraft which will land on the moon. Grumman Aircraft Engineering Corporation is prime contractor for the Lunar Module.

Bell tanks will be utilized in the lunar-landing spacecraft to supply water for the astronauts' consumption and for cooling the heat transfer section coolant of the LM during its descent flight to the lunar surface and ascent flight back to the orbiting Apollo spacecraft.

Positive expulsion systems are a necessity for space vehicles because liquids do not flow naturally toward a tank outlet as they would on earth. Instead, under the zero, or less than earth gravity conditions of space flight, the liquids tend to float in the tank, or cling to the tank walls.

The LM environmental control system water tankage for the ascent stage consists of two spherical, metal shell tanks containing a flexible, silicone rubber bladder and a perforated diffuser assembly, for positive expulsion of the water. The descent stage of the LM, used as the platform for liftoff, contains a conical, metal shell tank with the same positive expulsion water system.

When water is required, the pressurized gas collapses the bladder about the central perforated diffuser tube thereby forcing the water through the tube and out the outlet port of the tank. The bladder ensures delivery of the water to the environmental control system in all attitudes and gravitational fields.

177

3.1.6 Bell's Lunar Orbiter Propellant Tanks

3.1.6-a   Bell's Data Sheet for Lunar Orbiter propellant tanks

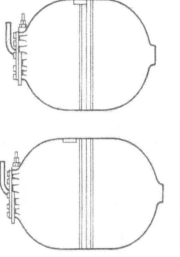

0   3   6   9   12   15
INCHES

FUEL TANKS:

PROPELLANT:	50/50 Blend Hydrazine Dimethylhydrazine
DIMENSIONS	
Diameter:	12.5 inches
Length:	17.3 inches
WALL THICKNESS:	0.027 inch
OPERATING PRESSURE:	190 psia
WEIGHT:	7.2 pounds
BLADDER MATERIAL:	Teflon
CYCLE LIFE:	20 expulsions
EXPULSION EFFICIENCY: 99%	

OXIDIZER TANKS:

PROPELLANT:	Nitrogen Tetroxide
DIMENSIONS	
Diameter:	12.5 inches
Length:	19.9 inches
WALL THICKNESS:	0.055 inch
OPERATING PRESSURE:	190 psia
WEIGHT:	7.9 pounds
BLADDER MATERIAL:	Teflon
CYCLE LIFE:	20 expulsions
EXPULSION EFFICIENCY: 99%	

The positive expulsion propellant tanks, which supply fuel and oxidizer to the rocket engine in the Lunar Orbiter spacecraft, are a product of Textron's Bell Aerosystems Company. They were designed, built and tested for The Boeing Company, principal contractor for the National Aeronautics & Space Administration on the Lunar Orbiter.

The 850-pound Orbiter vehicle was launched on its flight to the moon by an Atlas-Agena booster. Circling 25 miles above the lunar surface, the Orbiter is designed to take close-up photographs of possible landing sites for Apollo manned missions.

Bell Aerosystems also designed and built positive expulsion tanks for the Mercury spacecraft, Gemini-Agena Target Vehicle, Apollo Command. Service and Lunar Modules, Saturn S-IVB, Centaur, and USAF satellite program.

Positive expulsion is necessary because in the weightlessness of space propellants tend to float about inside a tank, or cling to its inner walls. They must be pushed out of the tanks and into the rockets.

Two cylindrical fuel and oxidizer tanks supply the propellants for the Lunar Orbiter. Running through the center of the cylinder, inside flexible Teflon bladder, is an aluminum diffuser tube, with more than 700 tiny holes. In filling a tank, the injected gas expands the bladder. The liquid-fuel or oxidizer is then fed into the bladder through the diffuser tube. As it enters the gas is bled out through a port.

When the rocket is to be fired, valves are opened between the tank and a nitrogen pressure tank. This gas compresses the bladder and the liquid is in turn delivered through the tiny holes into the diffuser tube and from the tube to the rocket.

The flow can be shut off by closing the gas valve. Pressure from the nitrogen gas remains on the outside of the bladder, and flow can be renewed instantly by reopening the valve.

178

3.2 Rocket Engines

3.2.1 Bell's Agena Rocket Engine

The Agena rocket engine was first used as the upper stage for the CORONA reconnaissance satellite. It was then utilized on numerous NASA and Air Force programs. Between January 1959 and February 1987, when the final launch occurred with a Titan 34B/Agena-D launch vehicle, 365 Agena engines were used. In the satellite programs, the payloads were mounted in front of the Agena engine and attached directly to mounting brackets on the engine frame; see Figure 3.2.1-e. The UDMH fuel and IRFNA oxidizer did not require an ignition system, as they would self ignite upon contact.

In the Gemini program, a Gemini spacecraft would chase and dock with a target vehicle. This allowed the astronauts to practice these maneuvers, in preparation for future space missions. The target vehicle was powered by a multiple start Agena engine. It turned out that it was necessary to restart the target vehicle's engine, on a number of missions.

3.2.1-a   Five Generations of Agena Rocket Engines

No.	Designation	Burn Time (sec)	Thrust (kilo Newton)	Application
A	8048	120	69	Upper stage Thor and Atlas rockets.
B	8081	240	71	Upper stage Thor and Atlas rockets. Launched SAMSOS, MIDAS Warning System, Ranger, Lunar Orbiter Lunar Probes.
C	Never built	---	---	---
D	8096	265	71	Upper stage of Thor, Atlas, and Titan rockets. Launched Gambit spy satellites, 3 Mariner probes to Mars. Used on Gemini target vehicle.
E	8247	Multiple start	71	Used on Gemini target vehicle.

Cenkner Table

Over the years, there were five major modifications to the basic Agena engine:

- A gimbal ring was added to the engine.
- The original JP-4 fuel was replaced by UDMH fuel.
- A restart capability was introduced, through the use of a propellant feed pump.
- Model 8096 was developed.
  - Added un-cooled nozzle extension.
  - Pump inducers were installed.
  - A high performance thrust chamber injector was added.
- Model 8247 was developed.
  - To implement a multiple engine restart capacity, liquid propellant rechargeable tanks were required.

The Apollo Lunar Module ascent engine was a modification of the Agena engine.

Reference

(1) *Agena: Enshrined in Smithsonian*, Bell Aerosystems

3.2.1-a   Bell's Data Sheet for Space Boosters utilizing bell's rocket propulsion systems

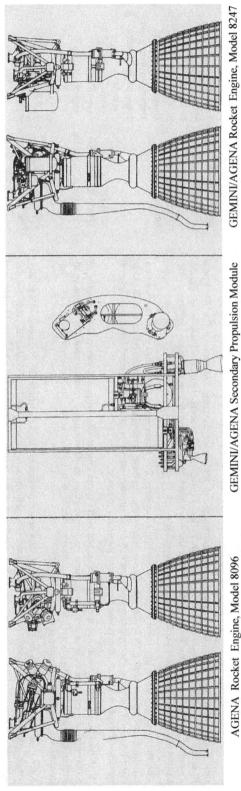

AGENA Rocket Engine, Model 8096

GEMINI/AGENA Secondary Propulsion Module

GEMINI/AGENA Rocket Engine, Model 8247

The highly reliable Model 8096 Agena Rocket Engine has been used extensively in USAF and NASA space projects. Project Gemini requires conversion to a multiple start rocket engine known as Gemini/Agena Model 8247. The Gemini/Agena Secondary Propulsion System affords maneuverability to the Agena Target Vehicle in the Gemini rendezvous missions.

SPACE BOOSTER MISSIONS

Thor/Agena	USAF Satellites	Atlas/Agena	USAF Satellite
	POGO		Eccentric Geophysical Observatory
	NIMBUS		Orbiting Astronomical Observatory
	ECHO		Advanced Passive Satellite
	TOPSIDE Ionosphere Sounder		Ranger
			Mariner
			Gemini/Agena
TAT/Agena	USAF Satellites		

180

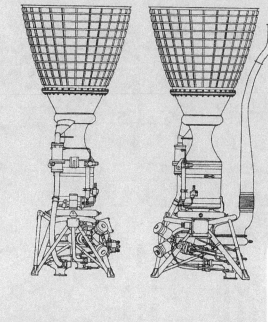

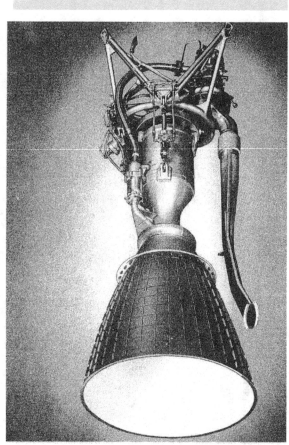

3.2.1-b   Bell's Data Sheet for Agena rocket engine, Model 8096

The Agena rocket engine, which powers the U.S. Air Force's Agena space vehicle, is known as "the work-horse of the space age".

Bell Aerosystems Company is associate prime contractor to the Space Systems Division of the Air Force Systems Command for the Agena engine. This liquid propellant rocket engine, which is designed, manufactured and tested by Bell, is shipped to the Lockheed Missiles and Space Company of Sunnyvale, California, prime contractor for the Agena space vehicle.

The Bell Agena engine has been fired in space approximately 200 times since 1959, and it has achieved a reliability record exceeding 99.4 percent. Agena space vehicles have orbited more than 80 percent of the U.S. Air Force and National Aeronautics and Space Administration satellites and have placed approximately 60 percent of the free world's functional unmanned payloads in space.

A 16,000-pound thrust liquid propellant rocket, the Bell Agena engine has propelled payloads in excess of 5,000 pounds into space. It is one of the first rocket engines to demonstrate restart capability. Continued improvement of this highly-reliable engine has resulted in a performance level higher than any other operational storable propellant rocket engine.

Agena currently is used in many USAF and NASA space programs. These include the Ranger and Mariner space probes to the Moon, Venus and Mars; the Nimbus weather satellite, Echo 2, the passive communications satellite; Alouette, the U.S.-Canadian topside sounder satellite, the Orbiting Geophysical Observatory, Polar Orbiting Geophysical Observatory, Advanced Orbiting Solar Observatory and the Orbiting Astronomical Observatory.

DIMENSIONS
Length:	Approx. 7 feet
Diameter:	35 inches
THRUST:	16,000 pounds
PROPELLANTS:	Unsymmetrical Dimethylhydrazine and Inhibited Red Fuming Nitric Acid
WEIGHT:	Approx. 290 lb
SPECIFIC IMPULSE:	Approx. 300 sec
CHAMBER PRESSURE:	Approx. 500 psi

 **BELL AEROSYSTEMS**
BUFFALO, NEW YORK – A **textron** COMPANY

181

3.2.1-c   Bell's Data Sheet for Agena Vehicle's secondary propulsion system

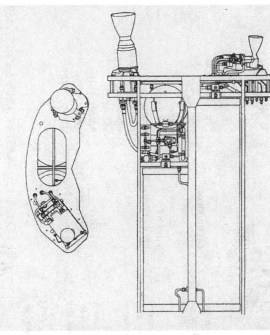

DIMENSIONS:
   Length: 51 inches
   Width: 35 inches
   Height: 15 inches
WEIGHT:
   Dry: 129.2 lb
   Gross: 303.8 lb
PROPELLANTS:
   Unsymmetrical Dimethylhydrazine and
   Mixed Oxides of Nitrogen
THRUST: 16 and 200 lbf

BELL AEROSYSTEMS
BUFFALO, NEW YORK – A textron COMPANY

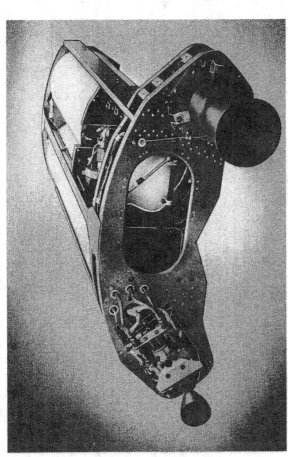

The Agena Target Vehicle's secondary propulsion system, designed and built by Bell Aerosystems, was used in space to reorient the propellants in the main tanks for each firing of the Bell Agena rocket engine. It also made vernier adjustments to the target vehicle velocity preparatory to the docking maneuvers of Project Gemini.

Each target vehicle had two secondary propulsion modules which fit snugly astride the Agena's aft rack. Each module consists of a set of 16 and 200-pound thrust, radiation-cooled rocket motors, positive expulsion propellant tanks, a pressurization system and various valves and controls.

The system is a storable liquid bipropellant system designed to supply thrust on demand for multiple firings under vacuum and zero gravity conditions. The rocket motors serve different purposes. The 16-pound tantalum tungsten motors are fired to force-feed the propellants from the tanks of the Gemini-Agena Target vehicle to the main propellant pumps so as to ignite the Agena engine.

In space, propellants, like other matter, tend to float or cling to the walls of the tanks. There is no gravity to hold them in the bottom of the tank. Therefore, acceleration of the vehicle is necessary to properly orient them. Propellant tanks for the secondary propulsion system itself use Bell-designed positive expulsion metal bellows which squeeze the propellant into the thrust chambers. The 200-pound rocket motors are fired to make fine adjustments to the velocity of the Target Vehicle.

Since 1954 Bell has designed and built reaction control hardware for the X-1B, X-15, Mercury, Centaur, X-20Dyna Soar, Lunar Landing Research Vehicle and the Agena.

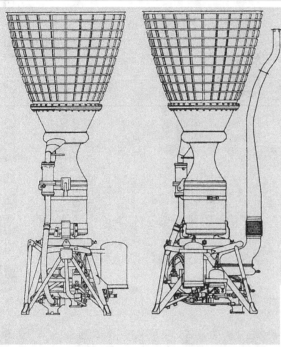

3.2.1-d Bell's Data Sheet for Agena rocket engine, Model 8247

The D.S_ Air Force Agena rocket engine, developed by Bell Aerosystems Company as the primary propulsion system for the Gemini Agena Target Vehicle, provides multiple start capability in space. Known as the Bell Model 8247 engine, it is a conversion of the standard Model 8096 Agen a which is capable of two starts in space.

To achieve this multiple start capability to meet the requirements of the National Aeronautics and Space Administration for Project Gemini, the Agena engine's starting system, fuel and oxidizer valves and electrical controls were changed. However, the turbine pump assembly and the thrust chamber were essentially left intact, thus carrying over the extremely high reliability of the earlier Agena engines.

The multiple start capability of the Gemini Agena engine fir st was proven in space during the Gemini 8 mission. Launched from Cape Kennedy on March 16, 1966, the Agena engine fired for about 186 seconds to inject the 3-1/2-ton target vehicle into a near-perfect 185-mile circular orbit for its historic rendezvous and docking with the Gemini 8 spacecraft.

In the three days following the docking and undocking, the Bell Agena engine was fired eight times on command from NASA ground controllers as they directed the target vehicle through a series of maneuvers in space.

These firings, which far surpassed the accomplishments of any other large rocket engine in space to date, were made to change the altitudes of the target vehicle's orbit and to change orbital angle to the equator (plane change). Duration of firings ranged from more than three minutes for the orbital insertion burn to a minimum impulse bit of less than two seconds for an orbit adjustment.

DIMENSIONS	
Length:	Approx. 7 feet
Diameter:	35 inches
THRUST:	16,000 pounds
PROPELLANTS:	Unsymmetrical Dimethylhydrazine and Inhibited Red Fuming Nitric Acid
WEIGHT:	Approx. 310 lb
SPECIFIC IMPULSE:	Approx. 300 sec
CHAMBER PRESSURE:	Approx. 500 psi

 BELL AEROSYSTEMS
BUFFALO, NEW YORK – A textron COMPANY

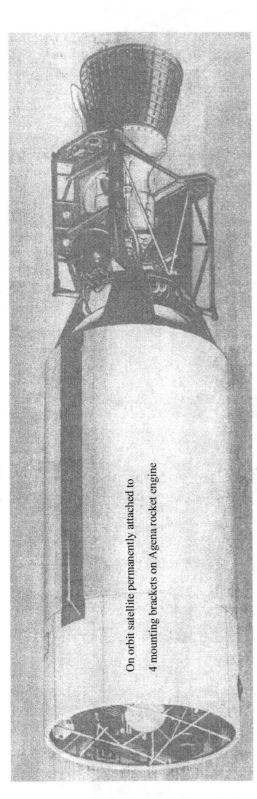

3.2.1-e Standard Agena space vehicle (diameter: 60 inches: length: 248 inches)

On orbit satellite permanently attached to
4 mounting brackets on Agena rocket engine

3.2.1-f Close-up Bell's Agena rocket engine

4 satellite
mounting
brackets

AGENA WORLD SPACE FIRSTS

- Rendezvous and docking of two spacecraft
- Achieved a circular orbit
- Achieved a polar orbit
- Achieved stabilization in all three axis in orbit
- Controlled on orbit by ground command
- Returned a man-maid object from space
- Propelled itself from one orbit to another
- Propelled a spacecraft on a successful Mars flyby
- Propelled a spacecraft on a successful Venus flyby
- Provided propulsive power for another space vehicle
- Propelled man to a new world altitude record: 852 miles
- Propelled man to a new world speed record: 17,930 mph

184

3.2.1-g   Schematic of Bell's rocket engine

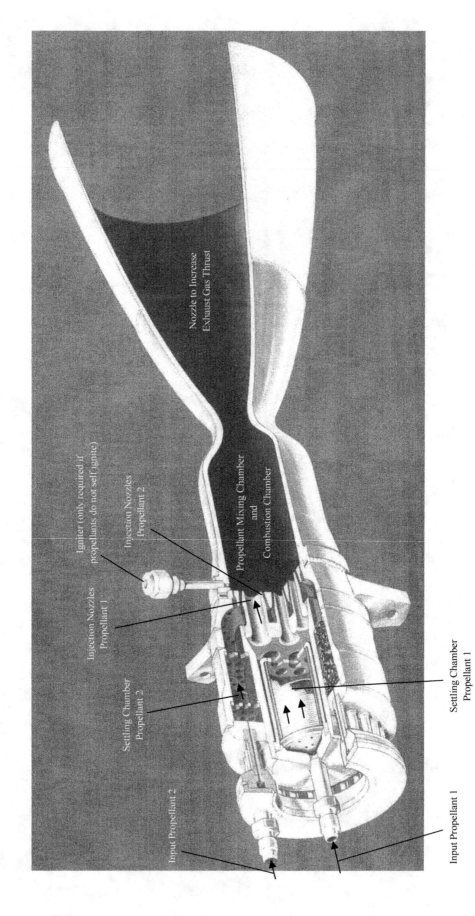

Igniter (only required if propellants do not self ignite)

Injection Nozzles Propellant 2

Injection Nozzles Propellant 1

Nozzle to Increase Exhaust Gas Thrust

Propellant Mixing Chamber and Combustion Chamber

Settling Chamber Propellant 2

Settling Chamber Propellant 1

Input Propellant 2

Input Propellant 1

3.2.1-h   Three rocket engines being tested simultaneously at the Wheatfield Plant

186

3.2.2 Bell's Lunar Module Ascent Engine

3.2.2-a Bell's Data Sheet for Lunar Module ascent engine

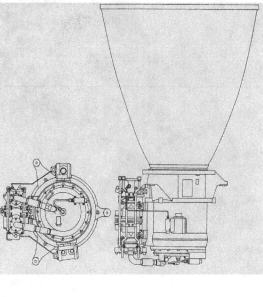

DIMENSIONS	
Length:	51 inches
Diameter:	32.6 inches (nozzle exit)
THRUST:	3,500 pounds
WEIGHT:	203 lb
PROPELLANT FLOW RATE:	
Fuel:	4.3 pounds/second
Oxidizer:	7.1 pounds/second
CHAMBER PRESSURE:	120 psi
EXPANSION RATIO:	45.6 to 1
RESTART CAPABILITY:	35 times
ENGINE LIFE:	460 seconds

BELL AEROSYSTEMS
BUFFALO, NEW YORK – A textron COMPANY

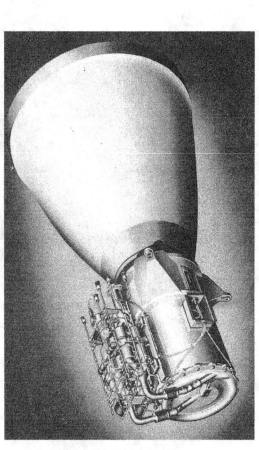

Liftoff of the National Aeronautics and Space Administration's Lunar Module from the lunar surface will be accomplished by means of the 3,500-pound-thrust ascent rocket engine, developed by Bell Aerosystems Company under subcontract from Grumman Aircraft Engineering Corp.

The Bell Ascent Rocket Engine generates the necessary thrust to lift the LM, less the expended descent stage, 50,000 feet into a lunar orbit for rendezvous with the orbiting Apollo spacecraft. The engine provides the necessary impulse for the gross orbit adjustment of the critical dock/mate maneuver. Once both vehicles are connected, the astronauts transfer from the LM into the Apollo Command Module. LM is jettisoned from Apollo to remain in lunar orbit, as the astronauts head Apollo for earth.

Bell's multiple restart rocket engine is a pressure-fed unit employing ablative thrust chamber cooling throughout, including the nozzle extension. The engine consists of a valve injector-trim orifice assembly, an ablative thrust chamber assembly with integral mounting features, electrical control harness, and instruments for measuring engine flight operation characteristics. Nitrogen tetroxide and a 50/50 blend of hydrazine and unsymmetrical dimethylhydrazine are used as propellants. The propellant valve incorporates a series-parallel redundancy start-stop capability into the engine. Ignition is initiated by hypergolic reaction of the propellants.

When the first NASA test firing of the Bell LM engine was conducted at the Manned Spacecraft Center's White Sands Missile Range Operation in New Mexico, NASA spokesmen termed the test, "Another major milestone in the United States manned lunar exploration program". Since then, more than 3,000 successful test firings have been achieved at sea level and simulated high altitudes.

187

3.2.3 Bell's Reaction Control Systems

Program		Agency Prime	Description	Function	Status	No. of Flights	Deliveries	Expulsion System	Propellants	Significant Accomplishments	Chamber Firings
X-1B 1954		NACA/ Bell Aircraft	Mono-propellant, 75 lb thrust	Attitude Control	Program Completed	2	4 Ship Sets (24 Thrust Chambers)	Piston Tank	H_2O_2	First Operational Reaction Control System	300
X-15 1956		NASA/ North American	Mono-Propellant, 43 & 113 lb thrust	Attitude Control	Program Completed	18	3 Ship Sets (36 Thrust Chambers)	Teflon Bladder	H_2O_2	First Reaction Control System Designed for Manned Aerospace Application	~6000
Mercury 1959		NASA/ Mc Donnell	Mono-propellant, 1,6, & 24 lb thrust, manual & auto	Attitude Control	Program Completed	12	25 Ship Sets (450 Thrust Chambers)	Silastic Bladder	H_2O_2	First Reaction Control System Designed for Manned Space Vehicle	—
Centaur 1960		NASA/ G.D. Astro	Mono-propellant, 1.5,3 & 50 lb thrust	Attitude Control, Propellant Orientation, Orbit Adjust	Program Completed	3	Thrust Chambers / Ship Sets 11 110 / 35 350	Silastic Bladder	H_2O_2	Improved Catalyst With Superior Low Temperature And Durability Characteristics	—
Agena		AF/ LMSC	Bipropellant, 16 & 200 lb thrust	Orbit Adjust And Propellant Orientation	Program Completed	—	19 Ship Sets (76 Thrust Chambers)	Teflon Bladder	Mon/UDMH	First Flight Proven Bipropellant Reaction Control System	5332
Gemini/ Agena Target Vehicle		AF/NASA/ LMSC	Bipropellant 16 & 200 lb thrust	Orbit Adjust And Propellant Orientation	Program Completed	—	—	Metallic Bladder	Mon/UDMH	First Multicycle Metallic Expulsion System & Modular Bipropellant	524
Maneuvering Satellite Reaction Controls		AF/ Bell Aerosystems	Gaseous Bipropellant	Attitude Control	Program Completed	—	—	—	OF_2/CH_4 And F_2/CH_4	—	—
X-20 (Dyna Soar)		AF/ Boeing	Mono-Propellant, 19, 34, & 42 lb thrust	Attitude Control	Program Completed	—	—	Silastic Bladder	H_2O_2	H_2O_2 System Operated In 1800 F Environment	7000
Lunar Landing Research Vehicles		NASA/ Bell Aerosystems	Mono-Propellant, 90 & 500 lb thrust	Attitude Control And Maneuvering	Program Completed	—	2	Gravity Feed	H_2O_2	First Free Flight Moon Landing Simulator	549
Reaction Control study		NASA/ Bell Aerosystems	Reaction Control State-of-the-Art Compilation	—	Program Completed	—	—	—	—	Comprehensive Bell Report Released No. 8214-933001	—

3.2.3-a   History of Bell Aerosystems reaction control systems

188

In space, to control the orientation of the vehicle and for fine adjustment of the spacecraft speed, reaction control systems are employed. Small rocket engines are mounted, at various locations on the vehicle that can be employed to adjust vehicle roll, pitch and yaw and to accelerate the vehicle, if required.

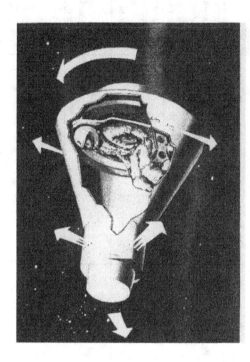

3.2.3-c Artists concept depicts possible Mercury spacecraft maneuvers using reaction control rockets.

3.2.3-b Artists concept showing how reaction control rockets will be used to maneuver Centaur spacecraft.

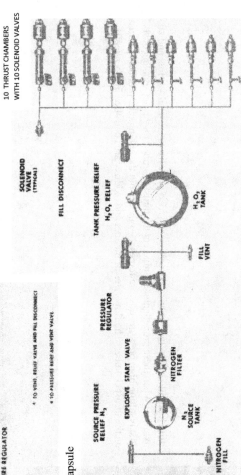

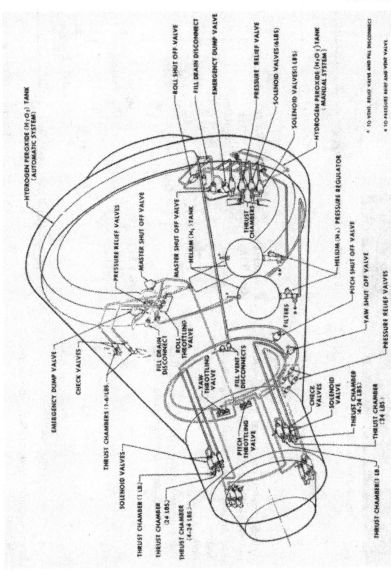

3.2.3-d   Liquid propellant reaction control system

Ten thrust chambers, with attached nozzles, are used to control the 3-dimensional orientation of the space capsule. The firing of each nozzle is independently controlled by a solenoid.

Prior to the mission, the evacuated propellant tank and N_2 tank are pressurized thru special fill vents.

Prior to changing the orientation of the capsule, the high pressure nitrogen gas is bled into the propellant tank to force the propellant to pressurize the feed lines to the thrust chambers; all solenoids are closed at this time.

To change the orientation of the craft, the astronaut manipulates the proper control to open the corresponding solenoid and fire the required thrust chamber.

The same basic technology was used with the Bell's rocket belt back pack; see section 2.5.1.

Reaction control system installed in space capsule

When hydrogen peroxide flows through a special porous catalyst bed, it decomposes into a 1360^0 F mixture of oxygen and steam. If this high pressure and high temperature mixture is then allowed to expand through a nozzle, thrust can be produced to propel the nozzle.

This principle is used in the reaction control system given in Figure 3.2.3-e. The liquid monopropellant H_2O_2 is stored in a spherical tank; pressure relief valves exhaust some of the propellant if the pressure gets excessively high. Nitrogen, stored in a spherical tank and connected to the H_2O_2 tank, is used to expel the propellant, on command.

3.2.3-e   Schematic of layout of space capsule reaction control system

3.2.3-f Liquid bipropellant reaction control system

With a liquid bipropellant reaction control system, two propellants are used – a fuel and an oxidizer. When these propellants are mixed in the thrust chamber, they react spontaneously. The resulting high pressure mixture forces the mixture out the exit nozzle, producing thrust on the thrust chamber. The flow rates of the fuel and oxidizer are controlled independently.

Installation in space craft

A high pressure gas, stored in a spherical tank, is used to force both propellants from their respective cylindrical storage tanks; Section 3.1.1 demonstrates how these tanks work.

The propellant valves are commanded on when thrust is required to reorient the vehicle.

Reference

Reaction Control Systems, Bell Aerosystems Company

16-Lb Thrust Chamber

200-Lb Thrust Chamber

Propellant Valve

Propellant Fill Valve

Oxidizer Tank

Tank Isolation Valve

Relief Valve

Gas Regulator

Check Valve

Filter

Gas Start Valve

Manual Gas Valve

Pressurizing Gas Tank

Gas Fill Check Valve

Fuel Tank

Propellant Fill Valve

3.2.3-g Schematic of layout of bipropellant reaction control system

191

3.2.4 Development of More Efficient Rocket Engine Propellants

For more than eight years, with company funding or under government contract, Bell worked on improving the efficiency of rocket engines by developing new propellants. These propellants would offer higher specific impulse (thrust per lb of propellant) that would increase the range of a craft, increase the size of the payload for a given mass of propellant, or increase mission flexibility. Two approaches were taken: replace the traditional liquid propellant with (1) fluorine or (2) fluidized small solid powders.

(1) Fluorine Propellant

The use of Fluorine as the oxidizer offers a number of attractive features:

(1) It will react with common fuels to generate the greatest possible thrust with significantly smaller storage tanks.

(2) Fluorine is self igniting when injected into the fuel. In contrast, systems utilizing oxygen and hydrogen initiate combustion with a separate ignition device.

(3) There is an abundant quantity of liquid Fluorine available.

(4) Allied Chemical Company, the largest producer of liquid Fluorine, invented a method for long term storage, and thus the shipping, of liquid Fluorine.

After selecting compatible materials, Bell developed a technique of passivation of the materials that come in contact with the fluorine. It became part of the normal cleaning process. This procedure places a thin layer of Fluorine on all contact surfaces. In 1956 the procedure was demonstrated by the successful firing of rocket engines with liquid fluorine. These engines ranged in size from 2,000 to 40,000 pounds thrust.

Besides using pure Fluorine, it was found that significant performance improvement could be achieved, with existing hardware, by mixing fluorine with liquid oxygen (FLOX). It was projected that, with a 30 % fluorine FLOX mixture, the Ranger lunar payload could be increased by 55 % and the Mariner spacecraft payload could be increase by 80% -- when used in the Atlas booster of the Atlas-Agena launch vehicle.

It was felt that the development of fluorine propellant technology was complete and it was ready for space applications. It was also felt that the most efficient launch vehicle would be achieved when fluorine was adapted for use in all stages of the launch vehicle, not just the upper stages – as was initially planned.

Reference

(1)Roach, R.D. Jr., *FLUORINE…QUEST FOR A BIGGER BOOST*, Bell Aerosystems, 1964

A fluidized powder is created when a gas passes through fine powder in a controlled way. The gas/powder mixture then behaves like a flowing liquid, and it can be controlled the same way a flowing liquid can be controlled, as the tables indicate, Bell funded laboratory testing revealed that fluidized powder propellants appear to promise advantages for long duration space missions that require long term storage with multiple restarts and throttle control.

The technology was successfully demonstrated on a rocket engine that generated 500 pound thrust.

Liquid Propellants
(two types: cryogenic and storable)

Higher impulse than solids
Multiple restart capability
Easily throttled
Cryogenic: highest impulse
        cold temperature storage
Storable: can be stored indefinitely

Fluidized Powder Propellants

Behaves like a flowing liquid
Multiple restart capability
Easily throttled
15% - 40% range increase over some liquids
Attractive for restart use when propellant is exposed to temperature extremes (i.e. space)

Solid Propellants

Lower impulse than liquids
Not easy to restart
Cannot be throttled
Easily stored for long periods
Less difficult to handle than cryogenic liquids
Less mission flexibility than liquids
Less expensive

References

(1) *Powdered Power*, Rendezvous, Bell Aerosystems, Vol. XI Spring 1972.
(2) *NASA, EXPLORING IN AEROSPACE ROCKETRY*, U.S. Government Printing Office, Stock No. 3300-0394, 1971.

3.3 Moon Vehicles

3.3.1 Bell's Lunar Landing Research Vehicle (LLRV)

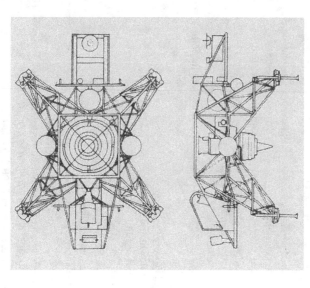

Height: 10 ft 6 in
Spread: Four truss legs spread 13 ft 4 in
Power Plants:
    Single gimbaled, vertically-mounted CF 700-2V axial flow
        aft turbofan engine (4,200 lbf thrust)
        Eight 500-lbf hydrogen peroxide lift rockets
        Sixteen reaction control hydrogen peroxide rockets
Gross Takeoff Weight: 3,710 lb
Maximum Altitude: 4,000 ft
Vertical Velocity: 100 ft/sec
Horizontal Velocity: 60 ft/sec
Capacity: Single pilot and 200 lb instrumentation

BELL AEROSYSTEMS
A Textron COMPANY
BUFFALO, NEW YORK

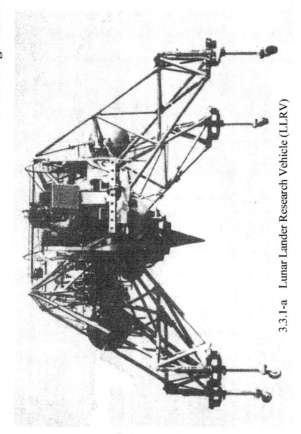

3.3.1-a Lunar Lander Research Vehicle (LLRV)

The Lunar Landing Research Vehicle (LLRV), designed by Bell for the National Aeronautics and Space Administration, is a non-aerodynamic, VTOL craft that was flown at NASA's Flight Research Center, Calif., to train pilots in lunar landing techniques -- here on earth.

As part of NASA's vast moon-bound Project Apollo, the LLRV employs a gimballed GE CF700-2V turbofan engine, mounted vertically in the center of the vehicle, to support 5/6 of the total weight -- simulating gravitational conditions on the moon. The remaining 1/6 of the weight is supported by means of lift rockets.

With this vehicle, a pilot can simulate in the atmosphere of earth, actual approach, hover, and touchdown procedures on the reduced 1/6 gravity moon. A variable stability autopilot system enables the pilot to achieve the same reactions and sensations as if he were operating in a lunar environment.

LLRV is designed so that various sections can be removed and replaced by actual hardware of the Lunar Excursion Module, the spacecraft in which astronauts will actually land on the moon by 1970.

Two LLRV's were delivered to NASA in mid-April 1964. NASA test pilot Joseph A. Walker made the first free flight of the LLRV at FRC on October 30, 1964.

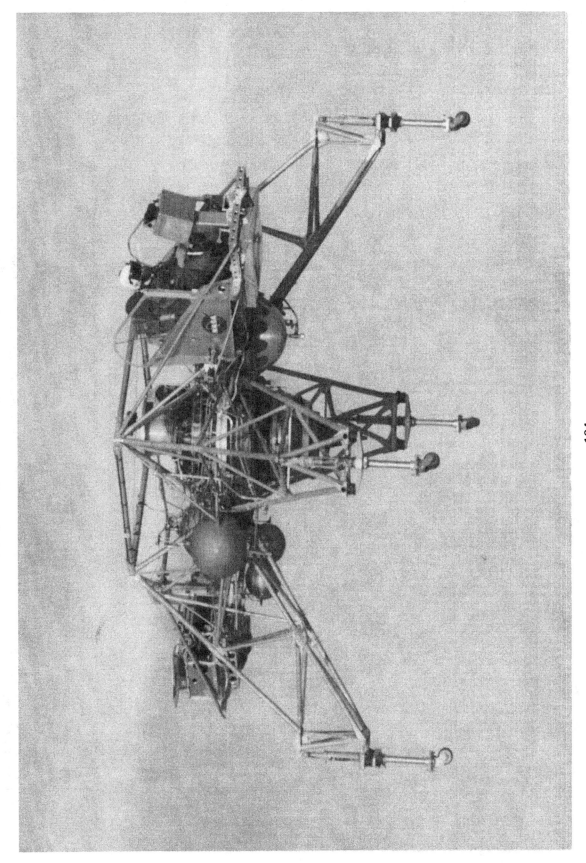

3.3.1-b   Astronaut practicing simulated landing on the moon using Bell's LLRV

194

3.3.1-c Port side view of Bell's Lunar Landing Research Vehicle (LLRV)

Equipment storage compartment

Two helium tanks

Vertical shock absorber

Horizontal rubber shock mount

Landing wheels

Two lift rocket engines

Gimballed turbojet engine

Six emergency rocket engines

Pilot cockpit with emergency ejection seat

Gimballed turbojet engine:
Full power – to ascend to operating altitude
Partial power – reduce vehicle weight to that on moon (1/6 g)
Two lift rocket engines:
Control descent under simulation of vehicle weight on moon
Six emergency rocket engines:
For use if the jet engine failed
Two sets of eight rocket engines: to turn vehicle, pitch it up or down

Helium tanks: Stores helium for pressurizing the rocket system

Equipment storage compartment: Stores the variable stability system, a radar altimeter, a
Doppler radar, and an 80-channel PCM telemetry system

Vertical shock absorber: Air-oil shock struts that attenuate the vertical touchdown energy

Horizontal rubber shock mount: Absorb horizontal touchdown energy

Emergency ejection seat: Rocket propelled Weber ejection seat

195

3.3.1-d   Aft View of Bell's Lunar Landing Research Vehicle (LLRV)

Equipment storage compartment

Port and starboard propellant tanks for lift and attitude rockets

Simulation of Moon Landing

Vehicle ascends to landing height (1,000 to 4,000 ft), using full power of jet engine

Jet engine power is reduced, to simulate the vehicle weight on moon, so that it supports 5/6 vehicle weight

Two lift rockets support rest of vehicle weight

Vehicle descends, as power of two lift rockets is reduced

Rocket thrust is used to compensate for wind forces, in order to simulate moon vacuum, by tilting craft while holding jet engine vertical

Craft can be moved horizontally using rocket thrust   (Matranga, G. J. and Walker, J. A., 1965)

3.3.1-e   Views of Bell's partially assembled Lunar Landing Research Vehicle (LLRV)

Top view from forward

Side view from forward

Pilots cockpit

Equipment storage compartment

Jet engine

Port side view

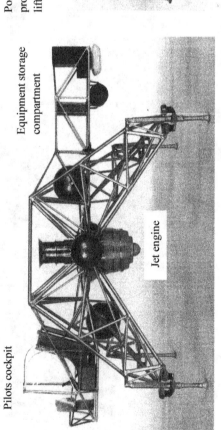

Port and starboard propellant tanks for lift and attitude rockets

Fore and aft jet fuel tanks for jet engine

Bottom view from forward

197

3.3.1-f Close-Up views of LLRV

Close-Up of Cockpit

Close-Up of Jet Engine

198

3.3.2 Bell's Lunar Landing Training Vehicle (LLTV)

3.3.2-a   Bell's Data Sheet for Lunar Landing Training Vehicle (LLTV)

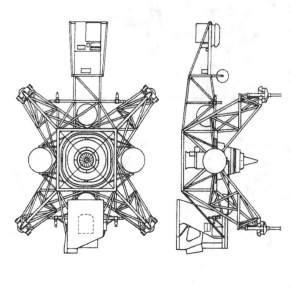

The Lunar Landing Training Vehicle (LLTV) designed and built by Bell Aerosystems for the National Aeronautics and Space Administration is an earth-based, non-aerodynamic craft being utilized to train astronauts in lunar landing techniques for the Apollo program.

The LLTV employs a gimbaled General Electric CF-700-2V turbofan engine mounted vertically in the center of the vehicle. The engine is constantly throttled to provide support for only five sixths of the total weight of the vehicle. Thus, the engine counteracts five-sixths of the earth's gravity. The remaining one-sixth earth gravity is comparable to the gravity of the moon. Lift for the remaining one-sixth of the LLTV's weight is provided by two Bell lift rockets. Controlled by the pilot, these rockets simulate the engine that is used for the lunar landing. A closed loop acceleration sensing system also automatically controls the jet engine attitude and jet throttle to cancel aerodynamic drag forces.

There are 16 attitude control rockets coupled with a variable stability autopilot system. Together these provide a simulation of the actual Lunar Module reaction control system. Settings of these rockets can be ground adjusted between 18 and 90 pounds thrust, permitting exact duplication of Lunar Module control characteristics.

Thus, as a training vehicle, the LLTV will permit the astronaut to practice approach, hover and touchdown techniques under the same gravitational and zero drag conditions he will find on the moon.

The LLTV is an extension of the Lunar Landing Research Vehicle (LLRV) concept pioneered by Bell Aerosystems in the 1962-64 period. Bell built two LLRVs which were flown in a research program at NASA's Flight Research Center, Edwards, Cal, beginning in 1964. These have since been converted to the LLTV configuration. Three new LLTVs were delivered by Bell to the NASA Manned Spacecraft Center, Houston, Texas, in the fall of 1967.

Height: 11 ft 4 in

Spread: Four truss legs spread 13 ft 4 in

Power Plants:

    Single gimbaled, vertically-mounted CF 700-2V axial flow aft turbofan engine (4,200 lbf thrust)

    Two 500-lbf hydrogen peroxide lift rockets

    Sixteen reaction control hydrogen peroxide rockets

Gross Takeoff Weight: 4,051 lb

*Maximum Altitude: 1,000 ft

*Vertical Velocity: 30 ft/sec

*Horizontal Velocity: 60 ft/sec

Capacity: Single pilot and 200 lb instrumentation

*For typical trajectory

3.3.3   Bell's Lunar Flying Vehicle (LFV)

3.3.3-a Bell's Data Sheet for Lunar Flying Vehicle (LFV)

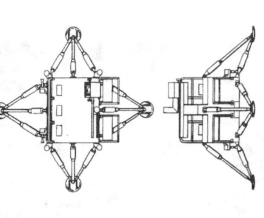

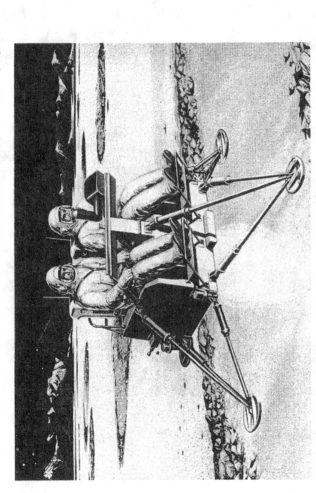

During 1964, space and lunar transportation engineers at Textron's Bell Aerosystems Company designed a versa tile, rocket-propelled flying vehicle for transporting astronaut-scientists over the lunar surface. The device, called a Lunar Flying Vehicle (LFV), is intended for operation in the vacuum and under the reduced gravity in the lunar environment.

The design concept of the Bell LFV is the result of a study conducted for NASA's Marshall Space Flight Center, Huntsville, Ala. The missions of the LFV are the return of astronaut-scientists to the Lunar Excursion Module (LEM), if a surface roving mobile laboratory (MOLAB) should be disabled and, to supplement the surface vehicle by permitting flight to areas which are inaccessible to a MOLAB.

The four-legged, 400-earth-pound LFV will be able to fly 50 miles non-stop. It is an open cockpit configuration, rectangular in shape, and is the size of an average desk. Astronauts equipped with backpack life support systems ride in the cockpit above the propulsion unit.

LEM propellants will be used for the five 100-pound thrust lift rockets, and the six 10-pound thrust rockets for attitude control. Bell has designed and delivered a Lunar Landing Research Vehicle (LLRV) to NASA for training pilots in lunar landing techniques he re on earth.

DIMENSIONS

Open:	92 x 92 x 67 inches
Folded:	38 x 60 x 61 inches

PROPELLANTS

Fuel:	50/50 blend hydrazine and unsymmetrical dimethylhydrazine (N2H4/UDMH)
Oxidizer:	Nitrogen tetroxide (N_2O_4)

WEIGHT

Dry mass:	Approx. 403 earth-lbs
Gross mass:	Approx. 1548 earth-lbs
Range:	50 miles
Crew:	Two astronauts

BELL AEROSYSTEMS

BUFFALO, NEW YORK   A textron COMPANY

200

3.3.3-b   Bell's Lunar Flying Vehicle (LFV) mock-up design tests with astronaut wearing space suit.

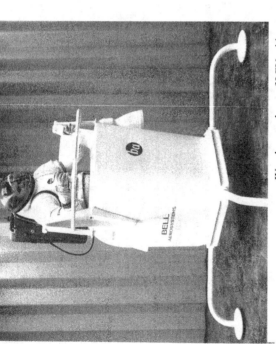
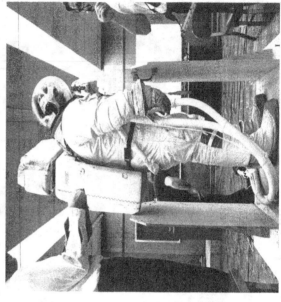

Wooden mock-up of LFV with astronaut in space suit operating controls.

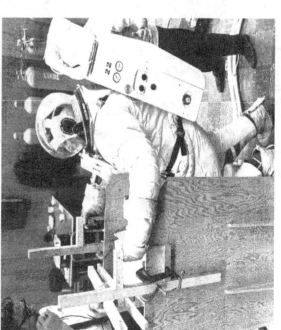

Wooden mock-up of LFV being built around astronaut wearing space suit.

3.3.4 Bell's Manned Flying System (MFS)

3.3.4-a Bell's Data Sheet for Manned Flying System (MFS)

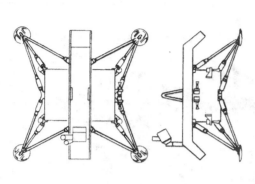

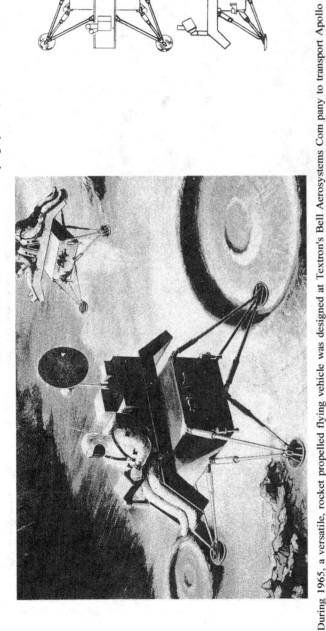

During 1965, a versatile, rocket propelled flying vehicle was designed at Textron's Bell Aerosystems Com pany to transport Apollo astronaut-scientists on exploration flights over the lunar surface. Known as a Manned Flying System (MFS), the device is designed for moon-based operations.

The design concept of the MFS, is the result of a follow-on to the earlier Lunar Flying Vehicle NASA Contract from Marshall Space Flight Center, Huntsville, Ala.

Primary mission for the MFS, is for 20 mile round trip, scientific explorations from the Apollo Lunar Excur sion Module (LEM). The MFS is capable of carrying one astronaut and 300-earth-pounds of scientific equipment to the bottom of rilles and craters, to the top of the scarps, and to ledges on steep slopes of the moon's surface.

The 400-earth-pound, four legged vehicle is capable of multi-hop flights, hovering in one area, or non-stop missions. It is an open cockpit configuration, rectangular in shape, and is the size of an average desk.

The major components of the propulsion system are the lift thruster assembly, six 10-pound attitude thruster assemblies, propellant tanks, and pressurization system.

Bell recently designed and delivered to NASA, a Lunar Landing Research Vehicle (LLRV) for training astronauts in lunar landin g techniques here on earth.

DIMENSIONS

Overall – less landing gear	Approx. 25 x 48 x 72 inches
Max. gear leg length	Approx. 59 inches

PROPELLANTS

Fuel:	50/50 blend hydrazine and unsymmetrical dimethylhydrazine (N2H4/UDMH)
Oxidizer:	Nitrogen tetroxide (N_2O_4)

WEIGHT

Dry mass:	Approx. 400 earth-lbs
Gross mass:	Approx. 1600 earth-lbs
Range:	50 miles one way or 20 mile round trip
Crew:	One astronaut
Payload:	300 earth-pounds

BELL AEROSYSTEMS

BUFFALO . NEW YORK   A textron COMPANY

202

3.4 Two-Stage Rocket-Booster/Glider–Space–Plane

3.4.1 Bell's Proposal for Development of Dynasoar (X-20) Glider–Space–Plane

3.4.1-a   Bell's two-stage rocket-booster/glider-space-plane -- as proposed for the Dyna-Soar Program -- mounted on Titan 2 booster launch vehicle. Intended applications included: reconnaissance, orbital bombing, orbital supply, satellite rendezvous and inspection, and research. The space-plane could orbit the earth, if required by the mission, and then glide down to a controlled landing. The space-plane was a predecessor to the space shuttle.

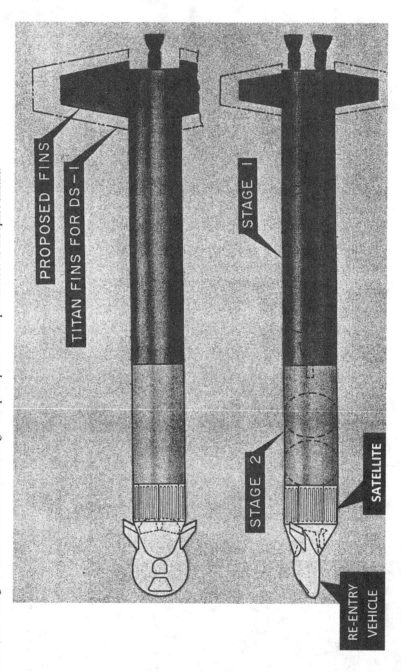

PROPOSED FINS

TITAN FINS FOR DS–I

STAGE I

STAGE 2

SATELLITE

RE-ENTRY VEHICLE

The glider-space-plane is mounted on a two stage Titan launch vehicle. To produce greater thrust, Stage 2 propellant is replaced by a fluorine-hydrogen propellant. The fins on the launch vehicle can be reduced in size, as shown, as a result of the size of the lenticular vehicle. If an emergency occurs during launch, or on the launch pad, the re-entry vehicle can be separated from the booster and launched to an altitude of 5000 feet, and then landed within seven miles of the launch site. The space-plane would be boosted, to hypersonic speeds, at an altitude of more than 30km, where it could glide between 10,000 and 41,000 km -- depending on mission requirements.

3.4.1-b Internal configuration of the satellite and the Lenticular re-entry vehicle

SATELLITE WITH RE-ENTRY VEHICLE ATTACHED

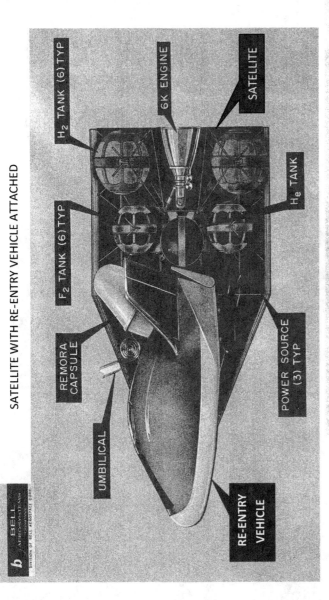

- H₂ TANK (6) TYP
- 6K ENGINE
- SATELLITE
- F₂ TANK (6) TYP
- He TANK
- REMORA CAPSULE
- UMBILICAL
- POWER SOURCE (3) TYP
- RE-ENTRY VEHICLE

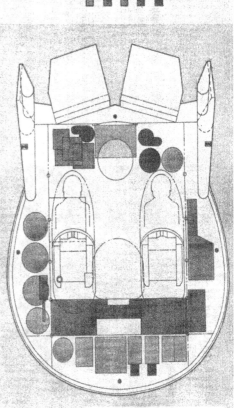

KEY

- ELECTRICAL
- GUIDANCE
- LIFE SUPPORT
- PROPULSION
- HYDRAULIC

3.4.1-c Re-entry vehicle to return two astronauts to earth from orbit

A shirt sleeve environment in maintained in the cabin during the entire mission. As an emergency backup, collapsible cocoons could be used to supply air for the astronauts. All equipment is located in the same cabin so that a single system provides environmental control for both equipment and astronauts. This arrangement also allows equipment repairs and maintenance to be easily performed by the astronauts.

3.4.1-d   Details on the Lenticular re-entry vehicle

THREE DIMENSIONAL VIEW OF LENTICULAR RE-ENTRY VEHICLE

ACCESS HATCH

ELEVONS

WHITE ABLATIVE COATING
ON BOTTOM OF CRAFT

WINDSHIELD

NOMINAL DIAMETER	11 FT
THICKNESS	5 FT
INTERNAL VOLUME	250 CU FT
RE-ENTRY WEIGHT	5783 LBS

Once the re-entry vehicle enters the atmosphere it can be controlled as a glider and landed on a conventional landing strip.

To protect the vertical fins from high angle of attack re-entry heating they fold inward, against the body, during re-entry. They fold outward forty five degrees, after reentry, to supply directional stability and longitudinal trim.

Longitudinal and lateral control is supplied by elevons.

White ablative material is coated on the bottom of the vehicle to protect it from re-entry heat. The vehicle enters the atmosphere with the bottom in the direction of travel, so that the bottom absorbs the re-entry heat.

LANDING PROFILE OF RE-ENTRY VEHICLE

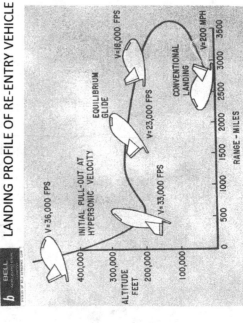

V=36,000 FPS

INITIAL PULL-OUT AT HYPERSONIC VELOCITY

EQUILIBRIUM GLIDE

V=18,000 FPS

V=23,000 FPS

V=33,000 FPS

CONVENTIONAL LANDING

V=200 MPH

ALTITUDE FEET

400,000
300,000
200,000
100,000

RANGE – MILES

500   1000   1500   2000   2500   3000   3500

THREE VIEWS OF LENTICULAR RE-ENTRY VEHICLE

The vehicle is comprised of a lens-shaped lower surface, in addition to a modified lens-shaped upper surface, with a rounded leading edge. It is a minimum size and minimum weight vehicle that can carry two or three astronauts.

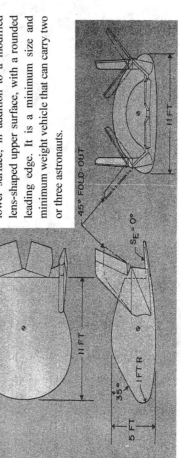

11 FT

11 FT

11 FT R

35°

5 FT

45° FOLD-OUT

$S_E = 0°$

205

3.4.1-e Bell Aerosystems proposal for demonstrating the feasibility of using a Lenticular Re-Entry Vehicle

1. Mission Analysis should be done to define the functions of the crew, to establish the equipment and onboard propulsion requirements, and to define maintenance and logistics requirements for various orbital missions. It is from these requirements that the configuration design and crew provisions are developed.

2. Configuration development to meet specific missions involves the preliminary design, technical analysis and detail installation drawings to show the feasibility and compatibility with the requirements. This work is necessary to determine the size, interior layout, and weight of the vehicle.

3. Wind tunnel tests should be conducted to obtain force and moment data at supersonic and hypersonic velocity. Instrumentation to obtain heat flux rates and temperatures is required for development of the heat protection system.

4. Development of combined radiative and ablative heat protection systems is required. This involves high-temperature materials with oxidation resistance, mechanical strength and insulation capability. Foamed and refractory oxides and carbides appear most promising.

5. A mockup is the next step in the development of a configuration. It demonstrates, in full scale, the feasibility of containing equipment, of accommodating a crew and of the ability of the crew to perform the functions of the mission. The human engineering aspects of cabin layout, visibility, operation of controls, and the psychological factors can be developed and evaluated.

6. An onboard propulsion system to perform the orbital missions, such as rendezvous and attitude control, requires a design analysis. This involves analysis of subsystems and the installation drawings, and the performance calculations to show that the system is compatible with the vehicle space and weight and suitable for the mission requirements such as storage time, weightlessness, and thrust control.

7. Flight testing of a Modified Lenticular Configuration can be started with Air Launch Tests of a glide vehicle. Dropped at Mach 0.7 at 40,000 feet from a B-52 would demonstrate the subsonic performance and handling qualities. The approach and landing techniques could also be developed. Such a vehicle need not have heat protection, propulsion or any of the long lead time subsystems required for space operation. Thus, it could be fabricated from conventional materials quickly for early testing.

References

(1) *DYNA-Soar: 1966*, Rendezvous, Bell Aerosystems, Vol II, N0. 5, 1963
(2) *Bell Modified Lenticular Re-Entry Vehicle*, Report No. D71110-953003, Bell Aerosystems, Sept. 1961.

3.4.2 Bell's Proposal for a Reconnaissance/Bomber Glider-Space-Plane (MX-2276)

Program Objectives

(A) System Requirements:

The goal of this activity is to develop a piloted hypersonic glider-plane system that can be used to obtain reconnaissance from various regions on earth or to accurately deliver a bomb to a designated target. The system is to be launched by a first stage rocket booster to a combat zone altitude between 100,000 and 150,000 feet. The basic mission distance can either be: (1) a mission radius between 1,500 and 2,000 nautical miles, or (2) a total range between 3,000 and 4,000 nautical miles. The hypersonic speed must be as high as possible without sacrificing altitude or radius.

(B) Reconnaissance Requirements:

In order of criticality, the acquired reconnaissance is to provide intelligence about:

(1) Strategic Warning: Determine if an enemy is planning an immediate attack as suggested by an assembly of forces and equipment and creation of new transportation systems.

(2) Operational Intelligence: Location, recognition, and identification of enemy troop concentrations or bases.

(3) Technical Intelligence: Identify the type of aggressive equipment the enemy posses.

(C) Reconnaissance Systems

(1) Photographic Reconnaissance: Two types of photographic missions are anticipated: search missions and detailed information missions. Each of these mission will have its own set of six dedicated cameras. The search mission will utilize two 27 inch, two 34 inch, and two 42 inch focal length cameras. The detailed information mission will require three 72 inch focal length cameras.

The target detail mission is intended to accumulate high resolution images of the target and the surrounding area. Detectable are 9 foot objects and recognizable are 25 foot objects, from an altitude of 150,000 feet.

(2) Radar Reconnaissance: Acquire meaningful information for present and future navigation, guidance, and bombardment systems. May be bombardment targets or navigation and guidance

(3) Ferret Reconnaissance: Monitor all radar frequencies that cannot be monitored from the ground. Acquire information about the source: transmission frequency, transmitter location, modulation type and characteristics, polarization, antenna pattern and scan rate. Create an area map showing locations of all radar sites and as much information as possible. Determine enemy's capabilities.

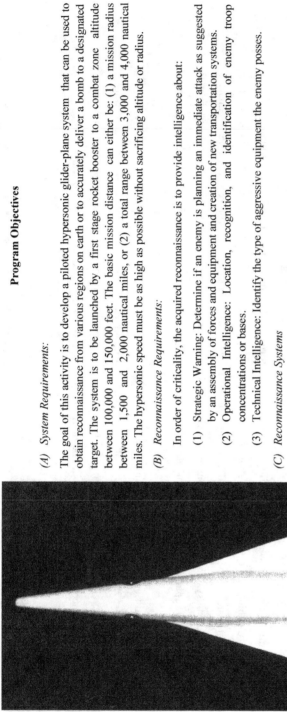
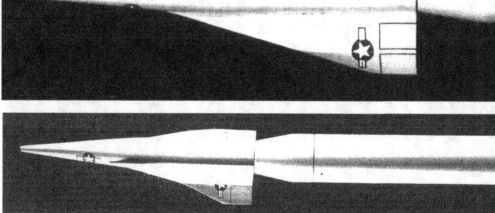
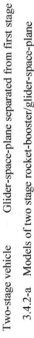
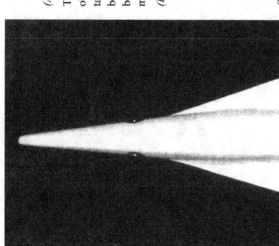

Two-stage vehicle   Glider-space-plane separated from first stage

3.4.2-a   Models of two stage rocket-booster/glider-space-plane

3.4.2-b   Two-stage launch vehicle for hypersonic glider-space-plane

115 ft. 8 in.

Gross Weight   142,460 lb
Propellants
JP-4   42,150 lb
Liquid Oxygen 84,300 lb
Liftoff Thrust   300,000 lb

41 ft.

143 in.

140 in.

10 ft

3.4.2-d   1ˢᵗ stage liquid propellant booster

3.4.2-c   2ⁿᵈ stage hypersonic glider-space-plane

System Design

(A) Selected Configuration

The selected vehicle configuration is a two-stage system. The first stage is an unmanned liquid rocket propelled booster attached, in tandem, to a second-stage liquid rocket propelled delta wing hypersonic glider-space-plane. The booster launches the system vertically to an altitude of 65,000 feet and a speed of 5,400 feet per second. At this point, the booster is jettisoned and the space-plane rockets are ignited, to increase its' speed to 6,600 feet per minute and it's altitude to 165,000 feet. From this point, the space-plane can achieve a range 4,640 nautical miles.

A delta wing design is employed for the airplane, with control being provided by a conventional rudder and an elevon. The first stage is controlled by gimbaled rockets

(B) Tandem Booster

A number of distinct advantages are offered by a tandem booster:

(1) Drag is minimized.

(2) As compared to a booster/airplane attachment to the side -- top or bottom -- separation is a lot easier.

(3) Physical attachment, with the airplane, is simpler.

(4) Alignment of drag, mass, and thrust, in a tandem configuration, is achieved.

(5) Problems associated with thrust, shear, bending, and torsional loads are minimized when large circular sections are joined.

(C) Stability and Control

Wind and gusts, during the first seconds after takeoff, must be compensated for to prevent the vehicle from becoming unstable. With the air speed being relatively low, fins are ineffective. Both rocket motors have therefore been gimbaled to give rapid roll, pitch, and yaw movement to provide stability and control of the vehicle.

(D) Heat Protection

During the mission both the booster and the airplane will experience aerodynamic heating. The booster will be protected by insulation while the airplane will be protected by a combine of insulation and the use of an active thermal control system. Most areas will be cooled by recirculated water while the wing leading edges will require a more sophisticated system – probably using lithium – due to the higher operational temperatures.

(E) Crew

The system will normally be operated automatically. Nevertheless, a crew is included to increase reliability and to include human capabilities that cannot be automated – assimilation and interpretation of various types of data and judgment to act intelligently upon the results.

Pilot

The pilot will have the responsibility of verifying that the plane has the proper angle of attack, velocity, altitude, etc at each point on the flight path. To achieve this, the pilot will correlate instrumentation reading with charts that show what this readings must be at each point on the flight path. If there is a discrepancy, she must identify the cause of the problem and fix it, if possible. A decision must then be made as to the best action to take: e.g. continue to fly the assigned path, switch to an alternate mission, return to home base, return to a backup base, or abandon the craft. The pilot will then manually fly the craft using whatever equipment is functional.

During the mission, the pilot will only have access to flight instrumentation information and information from the navigator – the periscope and radar will not be available. The periscope will be available during landing and the side hatches can be opened, to serve as windows.

Navigator

His primary responsibility will be to keep track of the current position of the craft, its' location in enemy territory, and the location of the return bases. If a control malfunction occurs, he must provide the pilot navigational assistance so the plane can return to base. He will also provided whatever reconnaissance support is required.

The navigator will have access to a visual display from the periscope, a radar display, and a map display. They will be used to verify that the plane has flown over various check points.

Reference

(1) Bell Aerospace Company, *MX2276 Reconnaissance Aircraft Weapons System Design*, Dec 01, 1955.

209

3.4.2-e   Configuration detail on hypersonic glider-space-plane. With the design range of 4,640 nautical miles, and the use of overseas takeoff and landing bases, this glider-plane will be able to monitor 80% of the key cities of the USSR. If the system is upgraded to a range of 5500 to 6000 miles, 100% of the USSR could be covered.

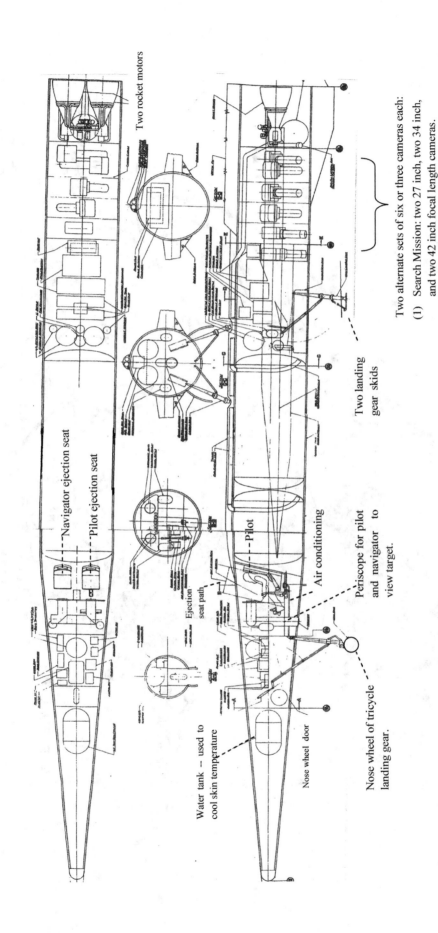

Two rocket motors

Navigator ejection seat

Pilot ejection seat

Ejection seat path

Pilot

Air conditioning

Periscope for pilot and navigator to view target.

Two landing gear skids

Two alternate sets of six or three cameras each:

(1) Search Mission: two 27 inch, two 34 inch, and two 42 inch focal length cameras.

(2) Detailed Information Mission: Three 72 inch focal length cameras.

Water tank -- used to cool skin temperature

Nose wheel door

Nose wheel of tricycle landing gear.

210

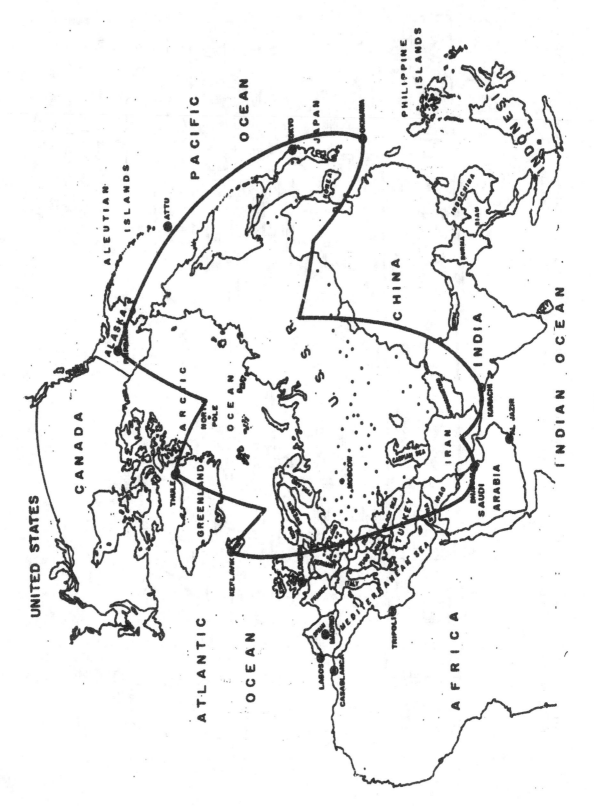

Figure 3.4.2-f   MX-2276 reconnaissance aircraft coverage with a 5500-mile glide path

211

3.5 Image Generators for Simulators

Generator Program Name	Image Generator Capability	Application
Agena Electronic Image Generator (AEIG) (3.5.1)	Computer generated image of Agena space craft as it rendezvous with another spacecraft.	Train astronauts in rendezvous and docking techniques. Simulation preparation for space rendezvous and docking of Agena spacecraft and target spacecraft.
Generator, Earth Orbit Scene (GEOS) (3.5.4)	Generates computer image of Agena target vehicle and earth as it rendezvous with Agena spacecraft.	Astronauts practice simulated rendezvous and docking.
Camera Mode Track (CMT) (3.5.2)	Camera translation track to generate TV image of terrain and rendezvous models for simulation.	Low level approximations for lunar surface, aircraft carrier for landing studies, airport for VTOL studies, and hydroskimmer base and launch ramp.
Electronic Planetarium (3.5.3)	Computer generated star fields.	Used with other programs: GEOS, RMU, and Agena target vehicle generator.
Remote Maneuvering Unit (RMU) (3.5.5)	Space vehicle image generator.	Studies related to control and retrieval of a remote space vehicle.
Spherical Model of Celestial Sphere (3.5.6)	Computer generated star images projected onto the inner surface of an opaque sphere.	Studies related to future space programs.
Manned Flying System (MFS) Simulator (3.5.7)	Utilizes: Camera Mode Track (CMT), Electronic planetarium, Remote Maneuvering Unit (RMU), TV Planetarium	Operator pilot gets: visual simulation of rocket flight over surface of moon, or space vehicle flying near mother ship, or aircraft landing on aircraft carrier, or vertical/short take-off and landings of aircraft.

Cenkner Table

212

3.5.1 Bell's Agena Image Generator ARIG)

3.5.1-a Bell's Data sheet for Agena image generator that was incorporated into the GEOS for rendezvous and docking training

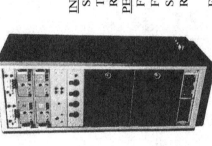

INPUT:	
Spacecraft	Roll, Pitch, Yaw
Target	Roll, Pitch, Yaw
Relative Distance	X/R, Y/R, Z/R, L/R
PERFORMANCE:	
Field Rate	60/sec
Frame Rate	30/sec
Scan Frequency	50kc
Resolution	0.2 inch on Vehicle Surface
Field of View	Changeable -- Originally 60°, now 90°
Range	10 feet to 50 miles
Target Vehicle Size	Length, 26 feet; Diameter 5 feet
Attitude Tracking Rate	30°/sec
Attitude Static Accuracy	6 min. of arc
Attitude Slew Rates	90°/sec
Attitude Acceleration Capability	30°/sec²

Agena vehicle simulated image

The Agena Electronic Image Generator was designed and built by Textron's Bell Aerosystems Company. It was delivered originally to NASA for incorporation in a Mercury Part Task Simulator, and was later incorporated into the GEOS (Generator, Earth Orbital Scene) equipment for the GMS (Gemini Mission Simulator) used to train astronauts in rendezvous and docking techniques.

A representative rendezvous image of the Agena is generated electronically and displayed through an oscilloscope and infinity optics system.

The target vehicle image configuration is that of a modified cylinder representing the Agena. Correct perspective size and angular orientation of this shape are simulated for ranges between the spacecraft and target vehicle from 50 miles to 10 feet. Markings are placed on one end of the generated image to aid the astronaut in determining target vehicle orientation and roll attitude during rendezvous and docking. The difference in the diameter of the ends also gives an orientation cue when the vehicle is at an angle with respect to the line-of-sight. Whenever the range is greater than 2000 feet, a flashing light indicates the position of the target. This is accomplished by a switching signal from the computer denoting that the image should be replaced with the flashing light at ranges of 2000 feet to 50 miles. Relatively simple computer programming is needed to drive the electronically generated image with respect to attitude and position. All window axis transformations and perspective computations are done within the image generator.

The major advantages of this device over camera model systems are its flexibility, full freedom, great dynamic range, space saving quality, and economy.

BELL AEROSYSTEMS A Textron COMPANY

BUFFALO, NEW YORK

213

3.5.2 Bell's Camera Mode Track (CMT)

3.5.2-a Bell's Data sheet for Camera Model Track (CMT) that simulates an astronauts view of the moon when landing

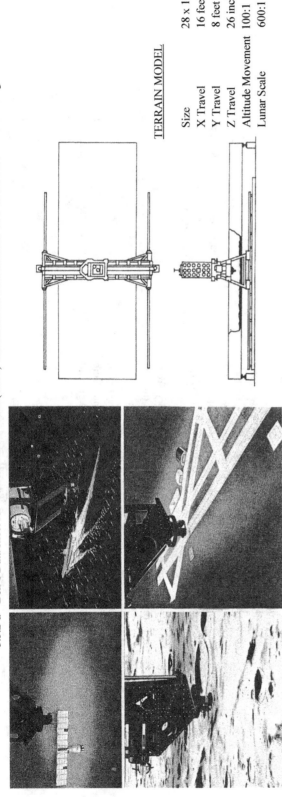

The CMT is a horizontally mounted gantry system designed and developed by Textron's Bell Aerosystems Company for the purpose of generating TV images of terrain and rendezvous models for visual simulation.

Six degrees of freedom and a number of terrain scenes give great flexibility for varied simulation problems. The terrain model initially installed in the system is a low altitude approximation of the lunar surface with realistically scaled relief features of craters, slopes, and ridges typical of those areas considered acceptable for landings. Mirrors are mounted on each side to extend the horizon.

Other models include an aircraft carrier and ocean scene used for carrier landing studies, an airport for use in VTOL contracts, a hydro skimmer base and launch ramp, and a variety of rendezvous models.

The models are installed near one end of the x-travel.

A model rotation servo is included to enable complete fly around without the use of Y travel. The TV camera and optical probe are driven longitudinally (X direction) on a gantry using four precision bearings, providing good low rate performance. Vertical travel (Z direction) is achieved through use of ball screws, and cross travel (Y direction) utilizing low friction recirculating ball linear bushings.

The servo driven optical probe embodies a highly precise mechanical design to ensure exceptionally smooth and reproducible motion in any mode. The optical probe entrance pupil can be positioned within 0.25 inch of the lunar surface.

TERRAIN MODEL

Size	28 x 10 feet
X Travel	16 feet
Y Travel	8 feet
Z Travel	26 inches
Altitude Movement	100:1
Lunar Scale	600:1

SERVO CHARACTERISTICS

X Maximum Rate	8 in. / sec
Y Maximum Rate	8 in. / sec
Z Maximum Rate	4 in. / sec
Low Rates	0.004 in. / sec
Accelerations	8 in. / sec^2

PROBE CAMERA

Pitch	360° Continuous
Azimuth	± 360°
Roll	360° Continuous
Look-Up Angle	45°
Field of View	60° to 30°, Can be Changed to 90°
Camera	GE TE-9
F Number	f 9.5
	T17
Focus Servo	∞ to 0.25 inch

BELL AEROSYSTEMS
BUFFALO, NEW YORK   A **textron** COMPANY

3.5.3 Bell's Electronic Planetarium

3.5.3-a Bell's Data Sheet for Electronic Planetarium that is used to generates background stars for use with other image generators

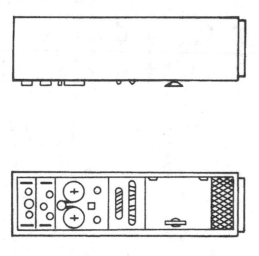

An Electronic Planetarium designed by Textron's Bell Aerosystems Company is a non-TV star generator intended for use with other electronic image generators such as GEOS (Generator, Earth Orbital Scene), the Agena target vehicle generator, or the RMU (Remote Maneuvering Unit) generator.

It uses the same coordinate transformation techniques as in GEOS, but has a core memory which is read out into the transformation at a 30/ second repetitive rate. The memory stores up to 512 star positions and magnitudes. The stars can be located anywhere in the entire celestial sphere.

The advantages of this device over the TV planetarium are that it has full 360⁰ continuous freedom in all three axes, and that the stars are not displayed on TV lines; therefore, very good low rate movement of displayed stars can be achieved. Window axis transformations are included in the generator. Oscilloscope type displays are used.

Using a 36-bit memory, stars can be digitally positioned to 1 in 1024 parts in each of the X, Y, and Z axes. An X, Y, Z rectangular coordinate system is used for star storage instead of the normal polar coordinator (declination, right ascension) for compatibility with the Bell GEOS type transformation system. By using 0.01 % digital-analog converters and 0.01% continuous transformation servos, star position accuracy approaching 0.01 % absolute is possible. Star magnitudes are stored in a 3-bit digital word, giving eight possible magnitudes.

The star information is automatically entered into the memory from one-inch punched tape during the 45 second system turn-on sequence.

Also since no TV raster lines are presented, stars may be located at any position on the display tube. This eliminates jitter and "line hopping" of stars. Star brightness with respect to background illumination is also greatly enhanced.

INPUT:

Spacecraft	Roll, Pitch, Yaw

PERFORMANCE:

Number of Stars	512
Position Accuracy in Memory	3.5 min. of arc
Magnitudes	any 8
Memory Type	Core
Frame Rate	30/sec
Attitude Freedom	360° Continuous All Axes
Slew Rates	90°/sec
Maximum Tracking Rate	30°/sec
Field of View	Changeable, Similar To GEOS

BELL AEROSYSTEMS A textron COMPANY

BUFFALO, NEW YORK

215

3.5.4　Bell's Generator, Earth Orbital Scene (GEOS)

3.5.4-a　Bell's Data Sheet for GEOS (Generator, Earth Orbital Scene), used to practice orbital docking of Agena and target spacecraft

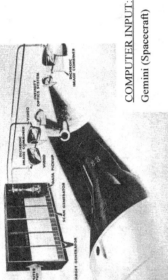

GEOS (Generator, Earth Orbital Scene) is a computerized visual scene generator, designed and built for the National Aeronautics and Space Administration by Textron's Bell Aerosystems Company. GEOS enabled Project Gemini astronauts to practice simulated rendez vous and docking procedures in preparation for actual space rendezvous missions.

The system produces a dynamic electronically generated color display which allows the astronauts to view, through their respective spacecraft windows, the earth and the Gemini –Agena Target Vehicle with its flashing light.

As the astronauts operate the controls to affect a visual rendezvous, the flashing light is first seen at 50 miles and continues to get brighter until the spacecraft is 2,000 feet from the target image. At that time, a picture of the Agena vehicle, in proper relation to the earth, is displayed to the astronauts.

The high resolution picture produced by Bell's GEOS provides images in the proper perspective, attitude and position in accordance with the control actions performed by the astronauts in the simulator. Using the displays, they are able to visually line up their spacecraft with the gradually approaching target vehicle until the final docking maneuver is accomplished.

A unique electronic image generation and scanning method is used instead of the traditional TV and gimbaled model techniques, allowing complete angular and positional freedom during rendezvous and docking maneuvers.

Earth data is obtained from a continuous 5-inch wide color negative film of the orbital track, and is presented on the displays through a high resolution video system, used simultaneously with a circular type scan.

COMPUTER INPUT:

Gemini (Spacecraft)	Roll, Pitch, Yaw
	Orbital Position
	Altitude
Agena (Target Vehicle)	Roll, Pitch, Yaw
	Displacement from Gemini
	X/R, Y/R, Z/R, 1/R

DISPLAYS:

21- Inch Oscilloscopes	One Pair per Window (Yellow and blue phosphors) optically mixed for color scene designed to mate with infinity optics system

PERFORMANCE:

Field Rate	60/sec
Frame Rate	30/sec
Type of Scan	
Target	50 kc,Image related
Earth	Spiral, 10 kc to 100 kc
Resolution	1500 Lines Horizon-to-Horizon
Brightness	200 Foot-Lamberts
Freedom	Full 6 Degrees of Both vehicles
Range	50 Miles to 10 Feet Continuously
Altitude	Variable 25 to 200 Miles
Video Bandpass	17.5 MHz

BELL AEROSYSTEMS

BUFFALO, NEW YORK　A **Textron** COMPANY

3.5.5 Bell's Remote Maneuvering Unit (RMU)

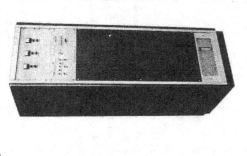

3.5.5-a   Bell's Data Sheet for RMU (Remote Maneuvering Unit) used to create image of target vehicle for Gemini rendezvous training

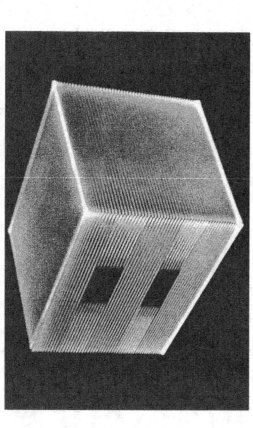

INPUT:
RMU   Roll, Pitch, Yaw
Relative Distance   X/R, Y/R, Z/R, 1/R

PERFORMANCE:
Field Rate   180 Hz
Frame Rate   45 Hz
Scan Frequency   11.5 kc
Vehicle Size   Each face variable up
                            to 40 in. x 40 in.
Resolution   120 line/face
Field of View   Changeable -- Presently 60⁰
Range   6 to 600 ft

The RMU (Remote Maneuvering Unit) Electronic Image Generator was designed and developed by Textron's Bell Aerosystems Company to perform in-house configuration studies related to the control and retrieval of a remote space vehicle.

It has the unique capability of presenting a stereo image from a 25 to 6-foot range, yet has a continuous range of greater than 100: **1.**

The image is generated entirely by a digital-analog device within the generator which has a field rate of 180 Hz and a frame rate of 45 Hz. Three fields are reserved for presenting the visible faces of the RMU box image. The fourth field is designated as a logic period. During this logic period, analog voltages are processed to make a decision as to which of the six sides of the box are visible to the viewer. This decision is stored for use during the following three frames. Only the visible sides of the RMU are generated by the electronics. Representative markings are provided on two opposite faces for additional attitude cues.

Video information derived from the processed analog signals is used to modulate the brightness of the individual sides of the RMU image as it rotates. Sides then appear or disappear without a sharp intensity change.

The attitude and position of the electronically generated image are computer-controlled by relatively simple programs. All window axis transformations and perspective computations are performed within the unit.

The image is displayed via two matched Schmidt oscilloscope projectors onto a 9 by 12-foot non-depolarizing rear projection screen.

BELL AEROSYSTEMS
BUFFALO. NEW YORK   A **textron** COMPANY

217

3.5.6 Bell's Opaque Spherical Model of the Celestial Sphere

3.5.6-a Bell's Data Sheet for 48-inch opaque spherical model of the celestial sphere, for simulated viewing of stars from a spacecraft

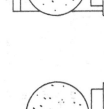

Two TV Planetariums of similar design and specification were developed by Textron's Bell Aerosystems Company. The first unit was delivered to Wright Field, Aerospace Medical Division in 1963, the second unit is being utilized at Bell Aerosystems Wheatfield Simulation Facility for studies related to future space programs.

The TV Planetarium is a 48-inch opaque fiberglass sphere with a high resolution television camera mounted within. The surface of the sphere contains 460 small holes at the locations of stars from the first to the fourth magnitude in the celestial sphere, the size of each hole being proportional to the star's brightness. The television camera images a spot of light transmitted through the star hole from the external illuminator, thus producing a picture of the star field.

The sphere is servo driven to rotate on a vertical axis to bring any desired hour angle of the celestial sphere into camera view. The camera is mounted on a gimbal whose servo driven axis is horizontal, allowing the camera to be positioned at the desired declination angle within the celestial sphere. Within the declination gimbal is a servo driven roll gimbal, permitting the camera to image the star field at the desired chase vehicle attitude. The sun and moon also can be installed as light sources on the star sphere.

The planetariums have been built with and without an internal servo driven earth occulting mask. When the earth occulting mask is not used, the star sphere can be installed in two positions oriented 90° from each other, so that orbits covering any portion of the star field can be simulated.

INPUT:		
Spacecraft		Yaw, Pitch, Roll
		Orbital Position
PERFORMANCE:		
Freedom		
	Outer Gimbal	360° continuous
	Second Gimbal	±80°
	Inner Gimbal	360° continuous
Maximum Rates,		
	All Axes	30°/sec
Maximum Acceleration		
	All Axis	30°/sec^2
Field of View		30° or 60°
Number of Stars	460	
Star Magnitudes	1 to 4	

BELL AEROSYSTEMS
A Textron COMPANY
BUFFALO, NEW YORK

3.5.7 Bell's Manned Flying System (MFS) Simulator

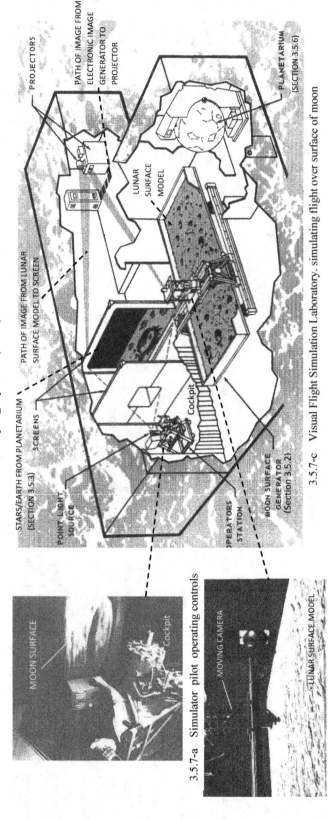

3.5.7-a Simulator pilot operating controls

3.5.7-b Television camera moving over model of lunar surface

3.5.7-c Visual Flight Simulation Laboratory. simulating flight over surface of moon

The Visual Flight Simulator could be used to simulate a number of situations: (1) A rocket vehicle flying near the surface of the moon, (2) A remote maneuvering unit (RMU) flying near a mother ship in space, (3) A plane landing on an aircraft carrier, (4) Aircraft vertical and short take-off and landings.

The above figures show a pilot simulating a flight near the surface of the moon. A camera that moves in three dimensions, moves over a model of the surface of the moon to simulate a vehicle maneuvering close to the surface. It can be adjusted to within ¼ inch of the surface, allowing the computer to conveniently make changes in scale and range, without changing the model. The cameras image is projected onto a large 10 x 14 foot screen, which has been positioned in front of the pilot. An electronic planetarium projects star images onto the screen background and onto a spherical planetarium in another room. Small scale lunar bases could be placed on the lunar model for practicing approaches. The pilot sits in a special cockpit, in a separate darkened room, where he can get a full view of the projected moon image. The computer synchronizes movement of the pilot's seat with the movement seen on the screen and it realistically controls instrument displays of velocity, altitude, attitude, time and fuel consumption.

To simulate the rendezvous of a robotic spacecraft with its mother ship, the arrangement in Figure 3.5.7-c is changed. The lunar surface model is replaced by a model of the mother ship. The moving TV camera now projects an image of the mother ship onto a TV monitor located in the cockpit. The electronic image generator projects a three dimensional image of the robotic spacecraft onto the large screen, observed by the pilot, while the electronic planetarium projects background earth and star images.

Bell created the first three dimensional simulator. This simulator could also create properly oriented shadows. This capability proved to be especially useful for studying vehicle behavior at close ranges.

Reference

(1) Eli, Perry, J., *Space Simulation*, Rendezvous, Bell Aerosystems, Vol. V, N0. 2, 1966

3.6 Space Electronics Systems

3.6.1 Bell's Radar Attitude Sensing System

3.6.1-a Bell's Data Sheet for the Radar Attitude Sensing System

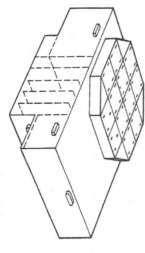

RASS (Radar Attitude Sensing System) is a lightweight, low power radar system, designed by Textron's Bell Aerosystems Company, which can determine the precise attitude, altitude, horizontal and vertical velocities of an orbital vehicle.

The radar uses sophisticated cw coding techniques to allow operation with low transmitted power. Verticality (pitch, roll) is determined by angle of arrival measurements on the return, using phase-monopulse techniques. Altitude is determined by measuring the delay of the coded cw return with respect to the transmitted signal, and velocities are obtained by Doppler techniques.

Bell is currently working under a NASA contract from the Manned Spacecraft Center, Houston, Texas, for system development. Qualification and orbital flight evaluation is scheduled for 1970.

DESIGN	
Volume	250 Cubic Inches
Weight	15 Pounds
Power	30 Watts

OPERATING FREQUENCY	X-Band

TRANSMITTED POWER	1 Watt

ANTENNA	Electronically steerable
	15 inch aperture

PERFORMANCE
(Based on average time of 1.0 second)

Roll	0.1 Degree
Pitch	0.1 Degree
Yaw	0.1 Degree
Altitude	150.0 Feet
Horizontal Velocity	0.2 Percent
Vertical Velocity	10.0 Feet per Second

BELL AEROSYSTEMS
BUFFALO, NEW YORK   A textron COMPANY

3.6.2 Bell's Satellite Communication Modem System

3.6.2-a Bell's Data Sheet for Satellite Communication Modem System

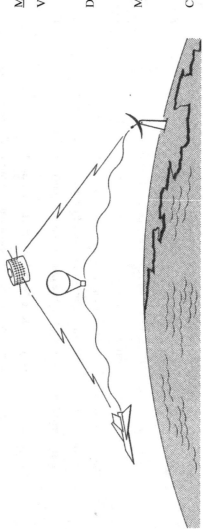

The Satellite Communications Modem System is designed for use in support of L-band satellite and balloon relay communications experiments. The system consists of voice and data modulation equipment for a ground transmitting station and receiving, demodulating and carrier-to-spectral noise density (C/N_o) measuring equipment for an airborne receiving station.

The modulator is a single rack mounting unit with a narrow band FM voice modulator and a differentially coded phase shift keying data modulator. The voice or the data may be selected to appropriately modulate the 70 mhz output carrier.

The receiving system has an L-band pre-amp followed by an L-band to VHF frequency translator. This permits the use of a standard ARINC VHF receiver, with the VHF receiver tuning control permitting L-band reception over the frequency range 1550 ±10 mhz.

The Demodulator and C/No Unit operate together to demodulate the 10 mhz I.F. output from the ARINC receiver and measure the incoming carrier-to-spectral noise density ratio (C/N_o) of the received signal. The demodulator will demodulate either the data or voice as modulated by the modulator unit.

MODULATOR UNIT

Voice Input	1 volt rms into 600 ohms 300 hz to 3500 hz bandwidths 6db per octave pre-emphasis
Data	Internal test data, 1200 or 2400 bits per sec. External data, 1200 or 2400 bits per sec
Modulation	
Voice:	FM, 1.5 khz deviation
Data:	DPSK, ± 90° deviation
Carrier Output	70 mhz; 0 dbm into 50 ohms
Frequency Stability	Better than 1 part in 10^6

FREQUENCY TRANSLATOR

Input Frequency	1550 mhz ± 10 mhz
Output Frequency	127 mhz ± 10 mhz
Frequency Stability	Better than 1 part in 10^6
Noise Figure	Less than 8 db
Gain	20 db

DEMODULATOR

Retrieval of voice and data inputs to Modulator Unit.

Data Error Rate within 1 db of theoretical.

Frequency Stability better than 1 part in 10^6

C/N_o MEASUREMENT UNIT

MEASUREMENT Accuracy Less than ± 1 db

3.7 Space Robot

3.7.1 Bell's Simulation of Remotely Controlled Space Robot

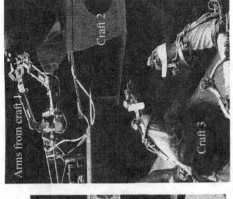

Simulated spacecraft 2
Moveable. Needs repair work.

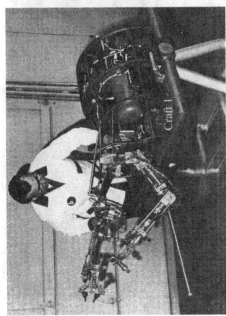

Simulated spacecraft 1
Mobile. Remote controlled
repair robot with two arms.

Simulated spacecraft 1
Mobile. Remote controlled
repair robot with two arms.

Simulated spacecraft 3
Stationary. Operator remotely controls
repair robot to replace parts on craft 2.

Perforated floor creates air cushion
to reduce drag on spacecraft 1 and 2
for more realistic space simulation.

Bell Aerospace was involved in the development of remotely controlled space robots since 1969. The space robot would be deployed and controlled by a mother ship. The robot would be guided to a spacecraft, in low earth orbit or in orbit around the moon or a planet, to perform any required assistance.

The space robot was to be used to:

(1) Retrieve orbiting satellites for repair or refurbishment.
(2) Undertake high risk missions without endangering the astronauts.
(3) Transfer cargo.
(4) Rescue astronauts, if required.
(5) Transport special instrumentation to remote areas.

To meet this goal, Bell designed and developed a space simulation test facility. Three simulated spacecraft were developed: two crafts (1 and 2) were movable by remote control and a third (craft 3) was stationary. Two articulating arms, on craft 1, were manipulated by an operator in stationary craft 3. Movement of crafts 1 and 2 were under the control of other operators (not shown) from individual consoles. These consoles had video displays from cameras mounted on the crafts.

To more realistically simulate movement in space, craft 1 moves on a cushion of air created by a 20 x 24 foot porous floor that is pressurized from below. Craft 1 can be controlled in forward and reverse directions, the lateral direction, and in yaw. Pitch and roll motion are controlled by installing the robot in a low friction gimbal ring. Only vertical motion cannot be simulated. The robot itself has five degrees of freedom.

In the above simulation, the robotic craft 1 has been moved to a disabled craft 2, for the purpose of removing disabled equipment. The operator in mother craft 3 manipulates the robot to replace the equipment.

The intent is to develop design criteria and explore operational characteristics to guide the development of an operational system.

References

(1) Fornoff, Heinz, *The Birth of the Mechanical Spaceman*, Rendezvous, Bell Aerospace, Vol. X/Spring 1971.

(2) Teleoperator: Part 2, Rendezvous Bell Aerospace, Vol Xii Spring 1973.

3.7.2 Bell's Concept for Manned Spaced Robot

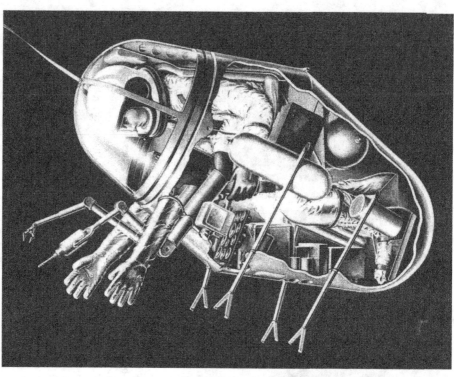

3.7.2-a Bell's manned space robot would be used to assemble, inspect, service, and maintain spacecraft. It would also allow astronauts to be transported between space vehicles. Grappling arms would attach the robot to the vehicle while the arms would be manipulated to perform assembly, service and maintenance.

3.8 Gemini Spacecraft

3.8.1 Gemini Simulates Apollo Docking

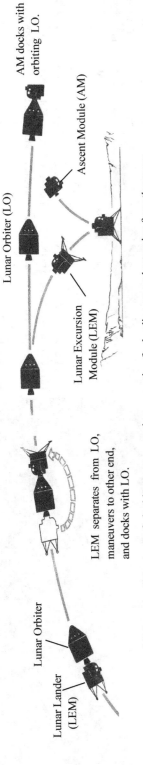

Lunar Orbiter (LO)

AM docks with orbiting LO.

Ascent Module (AM)

Lunar Excursion Module (LEM)

Lunar Orbiter

Lunar Lander (LEM)

LEM separates from LO, maneuvers to other end, and docks with LO.

3.8.1-a   Apollo spacecraft docking maneuvers in preparation for landing on and returning from the moon.

During the Apollo mission, to land humans on the moon, complex spacecraft rendezvous and docking procedures must be employed. First, the two separate sections of the Apollo spacecraft must separate and reorient themselves by employing a complex docking maneuver. The Lunar Lander separates from the Lunar Orbiter, and docks. Later on in the flight, the Lunar Lander disengages and lands on the moon. When the Ascent Module returns from the lunar surface, it must align itself with the Lunar Orbiter and dock.

To practice the complex alignment and docking procedure, and to prove that it can be done in space, with both spacecraft moving at 17,500 mph, the Gemini Program was created. Future missions, also requiring the capabilities developed by Gemini for use by Apollo, will no doubt include supply and crew transfer, a taxi to ferry personnel to orbiting space stations, approach and check out objects orbiting the station, and crew rescue.

In the Gemini Program, a separate Agena target vehicle, powered by Bell's Agena restartable rocket engine, would be launched into space, by an Atlas booster, to orbit earth. A two man Gemini space craft would then be launched, by a Titan II booster, to rendezvous and dock with the target vehicle.

The specific objectives of this program were:

"(1) Provide early rendezvous capability by (a) developing techniques, (b) assessing pilot functions, (c) developing propulsion, guidance, and control, (d) developing pilot disciplines, and (e) training pilots and providing them with rendezvous experience.

(2) Provide long-duration manned flight experience by (a) studying the effects of weightlessness, (b) determine physiological and psychological reactions to long duration missions, and (c) develop performance capabilities of the crew. "

The Gemini spacecraft will have safety ejection seats installed, much like those in a fighter aircraft. The astronauts can be safely ejected anywhere from the launch pad up to more than 60,000 feet.

Bell will build restartable Agena rocket motors for both the Agena-D Target Vehicle and the Gemini craft. A secondary propulsion system – comprised of two 16 pound and two 100 pound thrust rocket motors with Bell's positive expulsion storage tanks -- will also be built for providing fine control of the Gemini speed. The 16 pound rockets will also be employed to force feed propellants from the storage tanks to the main propellant pumps. A cold nitrogen gas reaction control system will rotate Gemini, when required. Bell's velocity meters will be installed to control engine shut off, once the vehicle has reached the targeted speed.

Prior to mission launch, the astronauts will have practiced extensively on a full mission Gemini trainer that is situated at Cape Kennedy. The simulator employs Bell's (GEOS, Section 3.5.4) software that creates images of the earth and the Agena-D Target Vehicle and allows the astronauts to practice docking.

Gemini steers toward Target Vehicle.

Astronaut spies Target Vehicle. At 20 miles, the astronauts will see high intensity flashing red lights on the Target Vehicle.

224

3.8.2   Gemini-10 Spacecraft Docks with Agena-D Target Vehicle

With a multi-restart Agena rocket engine, Gemini will have the ability to maneuver in space, at will; this includes changes in orbit. The 16 and 200 pound secondary thrusters will be used for fine adjustments during docking. Cold gas reaction control jets will be used to rotate Gemini, to get it oriented properly, just before docking.

Once docking orbit is achieved, the nose of Gemini must be rotated 180°, by the cold gas reaction control jets, so that it can mate properly.

One mockup and twelve operation craft, to be flown, were to be built. The first flight was to be an 18-orbit mission; the second was to last two weeks. With the fifth flight, rendezvous and docking would commence. The docking maneuvers will be executed by both astronauts.

During the Gemini-10 mission, the spacecraft remained docked to the target vehicle for 39 hours. The Gemini craft was maneuvered six times by firing the Agena engine and the secondary propulsion system. They first rendezvoused with their own target vehicle and then with Gemini-8 target vehicle. One astronaut was involved in two extra-vehicular activities and he maneuvered from one space craft to another and back. He also obtained a micro-meter experiment that was placed on the target vehicle four months earlier, before launch.

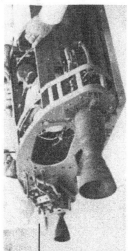

3.8.2-c   Single secondary propulsion system module. Each module includes a 16 lb and 200 lb thrust rocket motor, positive propellant expulsion tanks, a pressurization system, and support controls and valves.

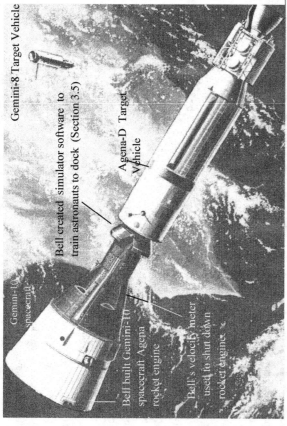

3.8.2-a   Gemini-10 spacecraft docked with Agena-D Target Vehicle

Gemini-8 Target Vehicle

Bell created simulator software to train astronauts to dock (Section 3.5)

Gemini-10 spacecraft

Agena-D Target Vehicle

Bell built Gemini-10 spacecraft Agena rocket engine

Bell's velocity meter used to shut down rocket engine

Bell built twin module secondary propulsion system

Flashing red lights for docking

Bell built Agena rocket engine for target vehicle

3.8.2-b   The Agena-D Target Vehicle is 32 feet long and 60 inches in diameter

References

(1) LEM – Excursion to the Moon, Rendezvous, Bell Aerosystem, Vol II, No. 4, 1963
(2) Gemini -- Next Step to the Moon, Bell Aerosystems, Vol II, No. 1, 1963.
(3) Spindler, A., It Was a Mighty Good Train, Rendezvous, Bell Aerosystems, Vol. V, No. 4, 1966.
(4) Mueller, G. E., Assessing or Manned Space Program, Rendezvous, Bell Aerosystems, Vol. IV, No. 5, 1965.

3.9   Space Shuttle

Bell conducted a study, in 1969, to identify ways in which its aerospace technologies and experience could be employed in the development of the space shuttle; four technologies were identified

AUXILIARY PROPULSION SYSTEMS

Technology	Application
Small rocket engines, controls and associated propellant tankage.	Maneuver and stabilize shuttle booster during launch. Maneuver and stabilize orbiter in space and during return to earth.
Utilize same propellants for the main engine and for auxiliary propulsion engines, thereby eliminating dual propellant storage and saving weight.	Liquid hydrogen and liquid oxygen propellants, for the main engine, are drawn off, converted to gas, and stored for use by auxiliary propulsion engines.
Advanced thrust chamber cooling techniques.	Engine can operate longer at higher temperatures with greater efficiency.
New propellant tank materials.	Increase life of propellant tanks by using long life corrosive resistant metallic bellows.
New graphite materials, which are 40 % lighter than aluminum but as strong as titanium.	Utilize graphite materials in propellant tanks and engine thrust chambers, to reduce weight.

AUTOMATIC APPROACH AND LANDING SYSTEMS

Technology	Application
Advanced navigational, targeting, and remote guidance systems.	Provide smooth landings for booster and orbiter, and for reentry of orbiter. System could handle launch/abort contingencies, in addition to emergencies related to manned/unmanned, powered/unpowered landings.

AIR CUSHION LANDING SYSTEMS

Technology	Application
Air cushion landing system to replace conventional landing gear.	Reduce weight, since it weighs less than conventional landing gear system. Can use a variety of different landing surfaces, which increases chances of survival with emergency landings. Can absorb greater landing shock loads and will distribute landing weight over larger area.

SPACECRAFT THERMAL PROTECTION

Technology	Application
Thermal control materials and systems to protest craft at hypersonic speeds and during reentry.	Drawing on extensive experience in the development of thermal control materials for hypersonic flight, special materials will be developed to meet the demands of the space shuttle mission.

Cenkner Tables

Reference

(1)   *Bell Aims Capabilities at Space Shuttle*, Rendezvous, Bell Aerospace, Vol. IX, No. 2, 1970.

4.0   Water/Land Craft

4.1 Development of Air Cushion Vehicles and Surface Effect Ships

In an air cushion vehicle the vehicle rides on a bubble of air trapped beneath the vehicle, which significantly reduces drag and increases speed. Shown below are basic approaches that can be used for creating this bubble. A fan is employed to force the air beneath the vehicle, and create the bubble, in sufficient quantity to support the vehicles weight. By attaching a flexible skirt around all four sides of the vehicle, the air leakage can be reduced and the vehicle will be able to move over uneven surfaces and over different types of surfaces.

A surface effect ship also rides on a cushion of air but the two sides have rigid skirts attached while the two ends still employ flexible skirts. The rigid side skirts remain in the water at all times so the vessel cannot move from water to land, like the ACV can. The SES is more resistant to sideward movement due to sideward sea or air forces and it could employ propulsion water jets since part of the vessel is always submerged.

4.1-a Basic Types of Air Cushion Vehicles

The open plenum concept is basically an inverted chamber with a hole cut in the top. Air pumped through this hole into the chamber creates a pressure sufficient to lift the craft. Daylight clearance is approximately equal to the thickness of the air jet which flows out of the bottom perimeter of the craft.

The annular jet, the most widely used means of developing an air cushion, attains a daylight clearance two to three times higher than the thickness of the escaping air. This is accomplished by fans forcing the air downward through slots around the periphery and slanting the air flow inward toward the center of the ACV. The angled air flow provides a seal that keeps about 60 percent of the air from leaking out - -thus providing the cushion with reduced power.

The addition of flexible trunks, or skirts, to an air cushion vehicle increases the obstacle or over-wave clearance by an amount almost equal to the skirt length, thus giving an ACV the performance capability of a much larger vehicle without skirts.

4.1-b Evolution of ACV "Skimmer" Technology at Bell

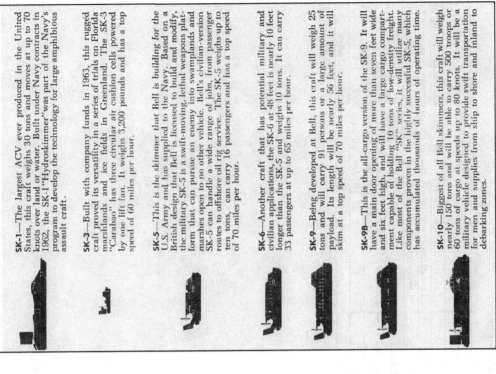

SK-1—The largest ACV ever produced in the United States, this craft weighs 30 tons and moves at up to 70 knots over land or water. Built under Navy contracts in 1962, the SK-1 "Hydroskimmer" was part of the Navy's program to develop the technology for large amphibious assault craft.

SK-3—Built with company funds in 1963, this rugged craft proved its versatility in a series of trials on Florida marshlands and ice fields in Greenland. The SK-3 "Carabao" has three circular air cushion cells powered by one lift fan. It weighs 3,200 pounds and has a top speed of 60 miles per hour.

SK-5—This is the skimmer that Bell is building for the U.S. Army and has supplied to the Navy. Based on a British design that Bell is licensed to build and modify, the military SK-5 is a swift-moving, lethal weapons platform that can pursue an enemy into swamplands and marshes open to no other vehicle. Bell's civilian-version SK-5 can handle a wide range of jobs, from passenger routes to offshore oil rig service. The SK-5 weighs up to ten tons, can carry 16 passengers and has a top speed of 70 miles per hour.

SK-6—Another craft that has potential military and civilian applications, the SK-6 at 48 feet is nearly 10 feet longer than the SK-5 and weighs 10 tons. It can carry 33 passengers at up to 65 miles per hour.

SK-9—Being developed at Bell, this craft will weigh 25 tons and will carry 91 persons or a large amount of payload. Its length will be nearly 56 feet, and it will skim at a top speed of 70 miles per hour.

SK-9B—This is the all-cargo version of the SK-9. It will have a main door opening of more than seven feet wide and six feet high, and will have a large cargo compartment capable of holding 10 tons of low-density freight. Like most of the Bell "SK" series, it will utilize many components proven in the highly successful SK-5, which has accumulated thousands of hours of operating time.

SK-10—Biggest of all Bell skimmers, this craft will weigh nearly 150 tons and will be able to carry 500 troops or 60 tons of cargo at speeds up to 80 knots. It will be a military vehicle designed to provide swift transportation for men and supplies from ship to shore and inland to debarking zones.

References

(1) 17 Years of SEV's, Rendezvous, Bell Aerospace Textron, Vol. XIV, Spring 1975

(2) Speed at Sea, Rendezvous, Bell Aerospace Textron

4.1-c   Surface Effect Ships Built by Bell

Purchased By	Name of Vessel	Length	Application
U.S. Coast Guard Wses-2	Seahawk	110 ft. SES	Law enforcement. Interdiction of illegal substances entering by sea.
U.S. Coast Guard Wses-3	Shearwater	110 ft. SES	Law enforcement. Interdiction of illegal substances entering by sea.
U.S. Coast Guard Wses-4	Petrel	110 ft. SES	Law enforcement. Interdiction of illegal substances entering by sea.
U.S. Coast Guard Wses-1	SES-200	160 ft SES	Study operation of large length over beam design.
U.S. Navy	SES-100B	78 ft SES	Confirm feasibility of developing 300 ton prototype.
U.S. Army Corps of Engineers	Rodolf	48 ft SES	Perform high speed hydrographic surveys.
Middle East Company	Margaret Jill	110 ft SES	Offshore oil crew boat.
U.S. Navy	Minesweeper Hunter (MSH)	189 x 39 ft SES	Clear mines from ports and coastal waterways.

Cenkner Table

4.1-d Bell ACV'S and SEV's that were built (solid black) or proposed (outlined) : 1959 – 1975

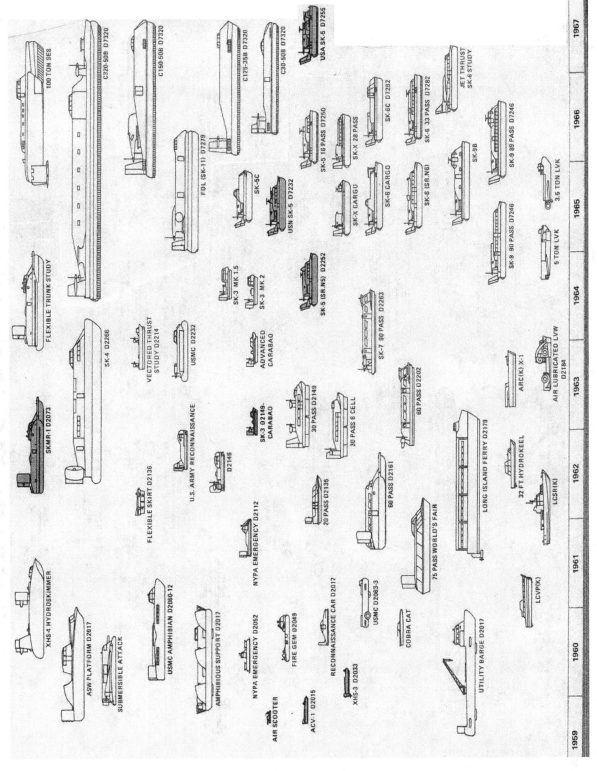

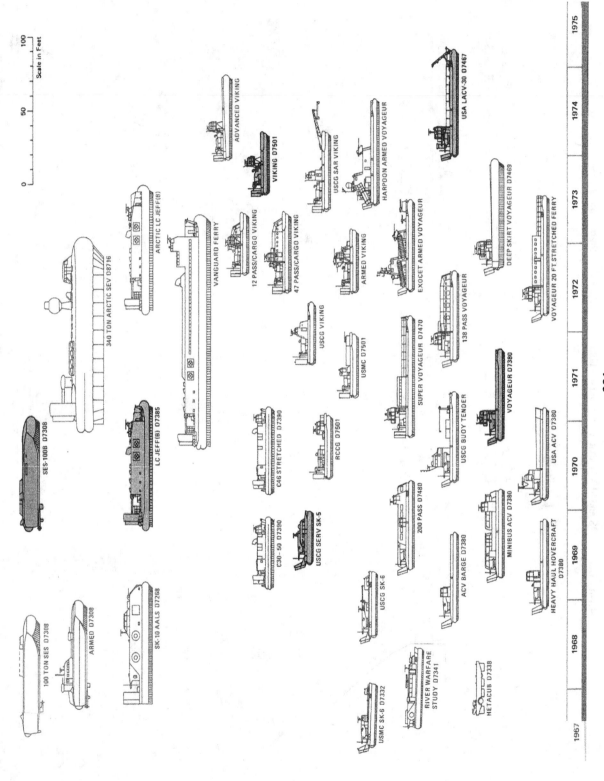

Scale in Feet

100 TON SES D7308
ARMED D7308
SK-10 AALS D7268
SES-100B D7308
340 TON ARCTIC SEV D8716
LC JEFF(B) D7385
ARCTIC LC JEFF(B)
VANGUARD FERRY
ADVANCED VIKING
12 PASS/CARGO VIKING
VIKING D7501
47 PASS/CARGO VIKING
USCG SAR VIKING
HARPOON ARMED VOYAGEUR
ARMED VIKING
USCG VIKING
EXOCET ARMED VOYAGEUR
USMC D7501
DEEP SKIRT VOYAGEUR D7469
C46 STRETCHED D7390
RCCG D7501
SUPER VOYAGEUR D7470
138 PASS VOYAGEUR
VOYAGEUR 20 FT STRETCHED FERRY
C30-50 D7390
USCG SERV SK-5
200 PASS D7480
USCG BUOY TENDER
VOYAGEUR D7380
USCG SK-6
ACV BARGE D7380
USA ACV D7380
USA LACV-30 D7467
MINIBUS ACV D7380
HEAVY HAUL HOVERCRAFT D7380
USMC SK-6 D7332
RIVER WARFARE STUDY D7341
HETACUB D7338

1967   1968   1969   1970   1971   1972   1973   1974   1975

231

4.1-e Bell Surface Effect Ships Design Proposals : 1959-1975

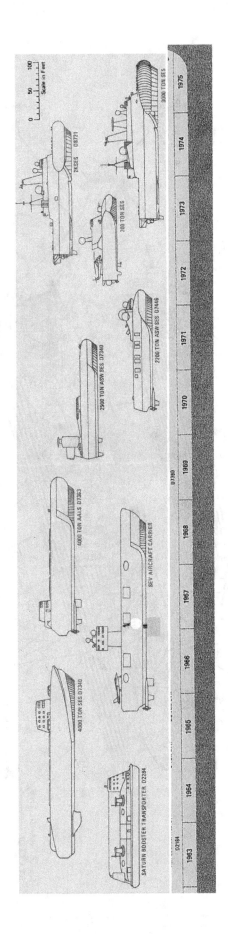

4.2   Details of Bell's LC JEFF(B) ACV

4.2-a   Drawing of LC JEFF(B) ACV showing craft layout

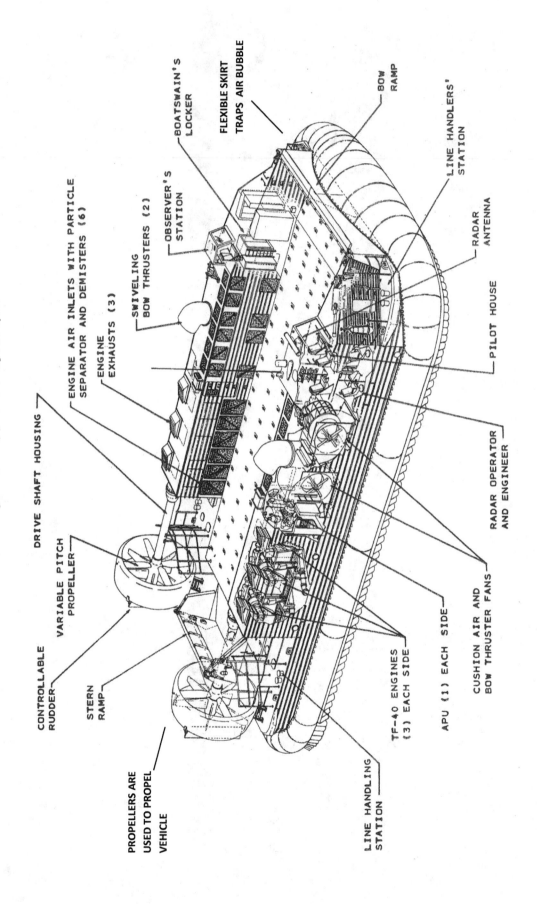

CONTROLLABLE RUDDER

DRIVE SHAFT HOUSING

VARIABLE PITCH PROPELLER

STERN RAMP

ENGINE AIR INLETS WITH PARTICLE SEPARATOR AND DEMISTERS (6)

ENGINE EXHAUSTS (3)

SWIVELING BOW THRUSTERS (2)

OBSERVER'S STATION

BOATSWAIN'S LOCKER

FLEXIBLE SKIRT TRAPS AIR BUBBLE

BOW RAMP

LINE HANDLERS' STATION

RADAR ANTENNA

PILOT HOUSE

RADAR OPERATOR AND ENGINEER

CUSHION AIR AND BOW THRUSTER FANS

APU (1) EACH SIDE

TF-40 ENGINES (3) EACH SIDE

LINE HANDLING STATION

PROPELLERS ARE USED TO PROPEL VEHICLE

4.2-b Assembly of Air Cushion Vehicles

234

4.3 Test Facilities for Small Scale Air Cushion Vehicle Models

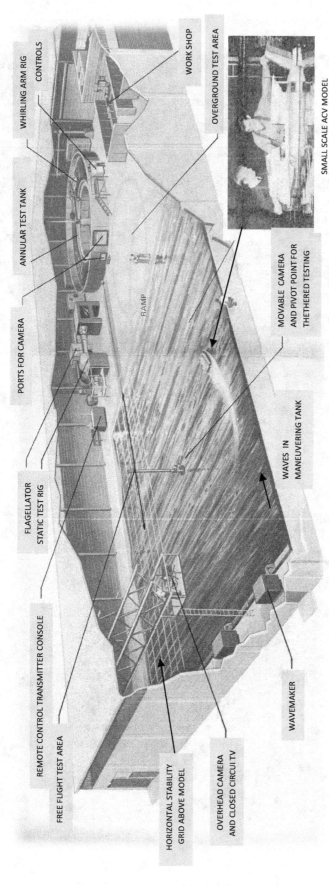

SMALL SCALE ACV MODEL

WHIRLING ARM RIG

CONTROLS

WORK SHOP

OVERGROUND TEST AREA

ANNULAR TEST TANK

PORTS FOR CAMERA

RAMP

MOVABLE CAMERA AND PIVOT POINT FOR THETHERED TESTING

FLAGELLATOR STATIC TEST RIG

WAVES IN MANEUVERING TANK

REMOTE CONTROL TRANSMITTER CONSOLE

FREE FLIGHT TEST AREA

HORIZONTAL STABILITY GRID ABOVE MODEL

OVERHEAD CAMERA AND CLOSED CIRCUI TV

WAVEMAKER

4.3-a   Air Cushion Vehicle laboratory for water tank testing of small scale vehicles

Annular Test Tank: The tank, which is 50 ft in diameter and 10 ft wide, holds 3 ft of water.

Whirling Arm Rig: A Model is mounted at the end of a 20 ft. arm, which is driven at speeds up to 40 ft/sec in the annular test tank, while drag measurements are made.; the model is free to pitch and heave. Overground testing can be conducted by placing a board over the water.

Maneuvering Tank (100ft x 100ftx 4 ft deep): Used for testing models under different conditions. A wavemaker, at one end, creates various types of controlled waves that propagate across the tank to a tapered ramp area, simulating waves washing up onto a landing beach.

Tethered Operation: A model that is driven by an electric or gasoline motor is tethered to the pivot point. The model is propeller in a circular path as measurem ents are taken.

Radio Controlled Operation: A gasoline motor provides power for both the lift fans and the propellers. Speed and direction are both controlled remotely.

Flagellator: Used to evaluate the performance of various skirt materials while they are exposed to sand, salt, and water under simulated ACV operation. Used to identify optimum skirt configurations and power requirements.

Stability Testing: For stability testing, a grid is placed above, or to the side of, the maneuvering tank. The performance of the model is recorded photographically while it is subjected to various speed and wave conditions. Radio controlled flights can be made at simulated speed s over 100 knots.

References

(1) ACV Technology Laboratory, Bell Aerospace Company
(2) Air Cushion Research, Bell Aerosystems Company

235

4.3-b Large Scale Voyageur Air Cushion Vehicle Being Prepared For Testing

4.4 Bell's SK-3 Carabao ACV

4.4-a Bell's Data Sheet for SK-3 Carabao ACV

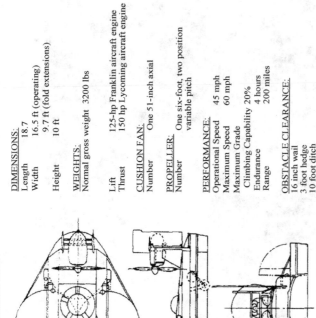

DIMENSIONS:
Length 18.7
Width 16.5 ft (operating)
9.7 ft (fold extensions)
Height 10 ft

WEIGHTS:
Normal gross weight 3200 lbs

Lift 125-hp Franklin aircraft engine
Thrust 150 hp Lycoming aircraft engine

CUSHION FAN:
Number One 51-inch axial

PROPELLER:
Number One six-foot, two position
variable pitch

PERFORMANCE:
Operational Speed 45 mph
Maximum Speed 60 mph
Maximum Grade
Climbing Capability 20%
Endurance 4 hours
Range 200 miles

OBSTACLE CLEARANCE:
16 inch wall
3 foot hedge
10 foot ditch

The Carabao is a rugged and versatile air cushion vehicle capable of performing a variety of missions over land, water, ice, snow, mud and marsh. The prototype vehicle began operational testing in March 1963.

During late 1963 and 1964, the Carabao completed a highly - successful series of operational demonstrations in the marshes and shallow waters of Lake Okeechobee, Florida on the James River at Fort Eustis, Va., and on the Potomac River at Washington, D.C. In July, the Carabao demonstrated the capability of the air cushion vehicle concept as a means of transportation in polar regions during a month-long evaluation program in Greenland.

The Carabao has three circular plenum cells, or air cushion, equally spaced around its center lift fan. Maneuverability is provided by rudders in the slipstream of its single variable pitch propulsion propeller. Two side wheels facilitate control over land and permit the machine to be towed along highways. The control system is designed to permit operations with little more training than normally required to operate an automobile.

An advanced version of the Carabao would carry a 1,200-pound useful load. The cabin is designed with maximum space and flexibility as prime considerations. For example, the new Carabao could carry a driver, three passengers and baggage, or a driver one medical attendant and two litter patients, or a driver and more than 1,000 pounds of cargo.

BELL AEROSYSTEMS

BUFFALO, NEW YORK   A textron COMPANY

237

4.5 Bell's SK-5 and SK-6 (stretched SK-5) ACV

4.5-a Bell's Data Sheet for military model of Bell's SK-5 ACV

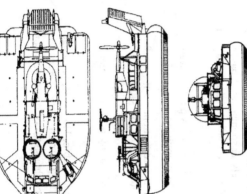

DIMENSIONS:
Length — 38 ft 10in
Beam (inflated trunks) — 23 ft 9in
Height — 15 ft 11in
Cabin floor area — 12 ft x 8 ft
Door opening — 5 ft 9 in high by 5 ft 7 in wide

WEIGHT:
Basic craft empty — 10,200lb
Gross weight (normal) — 17,000 lb
Additional load (overload) — 4,000lb
Gross weight (max. overload) — 21,000 lb
Passenger capacity — 16 plus operator

POWER PLANT:
Engine: One General Electric 7LM-100 PJ102 Marine gas turbine rated at 1150 shaft Horsepower at 80F
Propeller: One three-bladed, variable pitch, 9 ft Diameter Hamilton Standard
Lift Fan: One 17 ft diameter centrifugal
Fuel: Kerosene type JP4/JP5
Fuel Capacity: 304 gal

PERFORMANCE AT 1700 LB
Maximum speed: 60 knots (70 mph)
Range: 175 n mil at 50 knots
Endurance: 3.5 hr at 50 knots
Max gradient hover conditions: 1 in 7.5
Wave clearance at 40 knots: 4.5 ft

OBSTACLE CLEARANCE:
Solid wall: 3.5 ft   Earth mound: 5 ft
Vegetation: 5-6 ft   Ditch: 12 ft wide x 8 ft deep at 20 knots

The Bell SK-5 is a high-performance air cushion vehicle capable of a wide range of military applications on both land and over water. The SK-5 has served as both an assault and as a search and rescue craft. For high-speed personnel transportation, harbor and river patrol, and logistic support the SK-5 is equally adaptable.

The 7 1/2 ton SK-5 utilizes a flexible, air actuated skirt to travel on an air cushion more than four feet thick. Obstacle clearance and ditch crossing capabilities for the SK-5 are excellent.

In addition to the skirt-lifting system and the rudders mounted in the slipstream of the propellers, finger skirts and puff ports also aid in SK-5 control, being especially helpful in low-speed maneuverability. Flotation is assured by a large buoyancy chamber, comprised of watertight compartments, extending almost the entire length and width of the SK-5. In order to accommodate a jeep-size vehicle a cabin door approximately six feet by six feet has been provided. Hinged bow door windows and side access doors are also included.

The electrical system installation provides for easy access and maintenance. A self-contained auxiliary power unit for navigation and communication systems and a separate hydraulic system for controls activation are included.

Three SK-5s, tested at Bell's main plant and the Aberdeen Proving Ground, were accepted by the U.S. Army in April of 1968. Two are fitted as assault craft, the third as a transport. They are the Army's first ACV production buy, and the first combat ACVs built exclusively in the United States to military specifications.

BELL AEROSYSTEMS
BUFFALO, NEW YORK
A textron COMPANY

238

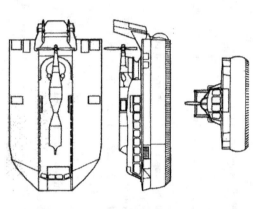

4.5-b   Bell's Data Sheet for the SK-6 ACV – the stretched version of SK-5

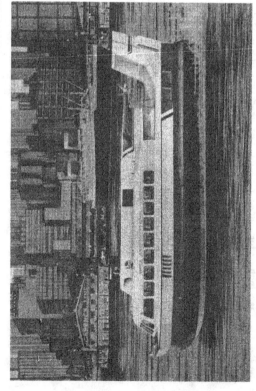

DIMENSIONS

Length	48 ft 6 in.
Beam (inflated trunks)	23 ft 9 in.
Height	15 ft 11 in.
Cabin floor area	21 ft 8 in. by 7 ft 8 in.
Door opening	5 ft 9 in. hight by 5 ft 7 in. wide

WEIGHTS

Basic craft empty		11,463 lb
Basic load		
crew ballast, fuel	2,718 lb	
Load allowance	7,819 lb	
	10,537 lb	
Gross weight (normal)		10,537 lb
Additional load (overload)		22,000 lb
		3,000 lb
Gross weight (max. overload)		25,000 lb
Passenger capacity		31 plus operator

POWER PLANT

Engine	One General Electric 7LM-100 PJ102 marine gas turbine rated at 1150 shaft-horsepower at 80°F
Propeller	One three-bladed, variable pitch, 9 ft diameter Hamilton Standard
Lift fan	One 7 ft diameter centrifugal
Fuel	Kerosene-type JP4/JP5
Fuel capacity	315 gal

PERFORMANCE: (at normal gross weight in I.S.A. conditions

Maximum speed	56 knots (66 mph)
Range	160 n.mi.
Maximum gradient at hover conditions	1 in 10
Maximum gradient (50 yd) Capability at 26 knots	1 in 3.5
Cruising speed in 4-5 foot waves	35-45 mph
Endurance	3.5 hr at 50 knots

OBSTACLE CLEARANCE

Solid wall	3 ft 6 in.
Earth mound	5 ft
Vegetation	6 ft
Ditches up to 16 ft wide and 8 ft deep can be crossed at 22 knots.	

BELL AEROSYSTEMS

BUFFALO, NEW YORK   A textron COMPANY

The Bell SK-6 Air Cushion Vehicle (ACV) is a versatile craft with extensive capabilities for military and commercial applications. Being completely amphibious it can operate from relatively unsophisticated bases for search and rescue missions, fire fighting, harbor and river patrol, for high-speed public transportation, and as a logistic support craft or utility freight carrier.

A single 1150 shaft-horsepower marine gas turbine engine drives both the lift fan, which forces air downward to create the air cushion beneath the craft, and the aft mounted propeller which provides propulsion.

Although the ten-ton SK-6 appears to ride a few inches above the surface, the hard bottom of the craft is supported by a four foot cushion of air. This is accomplished by flexible skirts, which give the SK-6 excellent obstacle clearance or ditch crossing capability over land and improved riding qualities over rough water and waves. Control of the SK-6 is achieved by a skirt-lifting system, a puff-port system and by a rudder mounted in the slipstream of the propeller.

The SK-6 accommodates 33 passengers in its roomy, soundproofed, air conditioned cabin. However, passenger seats can be quickly removed to convert the craft into a cargo carrier with 164 square feet of floor space. The SK-6 is a stretched version of the proven Bell SK-5 vehicle which has accumulated thousands of operating hours throughout the world.

4.5-c Operation of Bell's civilian passenger version of the SK-5 ACV

A Radar Compass
B Special Test Equipment
C Radio Microphone
D Oil Temperature Gauge
E Oil Pressure Gauge
F Engine Start Switch
G Oil Quantity Gouge
H Fuel Gauge
I Power Turbine Inlet
  Temperature Gauge

J Gas Generator Tachometer
K Power Turbine Tachometer
L Propeller Pitch indicator
M Air Speed Indicator
N Skirt Lift Control
O Radio Panel
P Ignition Test Switch
Q Elevator Trim Control
R Throttle and Pitch Control
S Rudder Control Pedals

In the SK-5 ACV, a single turbine engine is used to provide power to the lift fan and to the propulsion propeller.

The turning radius is much larger than that of a conventional boat, at high speed. At low speed, it can spin around in roughly the radius of the craft – about 40 feet for the SK-5.

If hit by a side wind, the vessel will move sideways at essentially the speed of the wind. To prevent this, the skirt must be raised on the appropriate side.

The normal procedure for stopping the vessel is to reduce power and reverse the propeller pitch. If the SK-5 is moving at 46 mph, it would stop in 150 yards – about twice the distance of a car on dry pavement. This stopping distance can be reduced to 50 yards, if the skirt is allowed to drag in the water.

Reference

(1) Helms, Daniel G., *How It Feels to Skipper a Skimmer*, Rendezvous, Bell Aerosystems, Vol. VI, No. 2, 1967.

Twist to change engine power.
Push to change propeller pitch +30° to -20°.

Hydraulic jacks lift the rubber skirt at two locations on each side, to control side force on the craft. The craft tilts when the skirt is raised on one side, and it moves sideways.

Foot pedals control the vehicles rudders for turning

4.6 Bell's SK-9 (passenger) and SK-9B (cargo) ACV

4.6-a   Bell's Data Sheet for SK-9 ACV

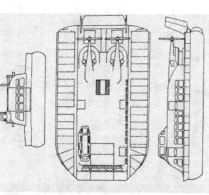

DIMENSIONS:

Length	55 ft 8 in
Width	32 ft 10 in
Height	16 ft 6in
Cabin floor area	29 ft x 17 ft 8 in
Door opening	7 ft 4 in high by 6 ft 4 in wide

WEIGHT:

Gross weight (normal)	47,000 lb
Additional load (overload)	52,000lb
Normal useful load	22,750 lb
Passenger capacity	90

POWER PLANT:

Engine: Two marine gas turbines rated at 1250 shaft horsepower

Propeller: Two four-bladed, variable pitch, 9 ft diameter

Lift Fan: Two 7 ft diameter centrifugal

Fuel: Kerosene

Fuel Capacity: 680 gal

PERFORMANCE (normal gross weight)

Maximum speed:	70 mph
Range:	210 miles
Max gradient -- static conditions:	12%
Cruising Speed in 4-5 foot waves	45 mph

OBSTACLE CLEARANCE:

Solid wall:	3.5 ft
Vegetation:	6 ft
Ditch:	18 ft wide x 8ft deep can be crossed at 25 mph
Earth mound:	5 ft

The high performance Bell SK-9 Air Cushion Vehicle is an amphibious transport designed for passenger - carrying operations. It will cruise over calm water at 65 MPH, and through 4- to 5-foot waves at 45 MPH with a full payload. A cargo transport version, the SK-9B, is also available.

The SK-9 utilizes many of the proven components of the highly successful SK-5 which has accumulated thousands of hours of operating time throughout the world in commercial and military operations.

The large cabin accommodates 90 passengers comfortably and has provisions for two tons of baggage or cargo. Passenger loading doors, which are located in the bow and on each side of the cabin area, allow quick loading and unloading to minimize turn-around time. This feature combined with the fast cruising speed of 65 MPH is essential to an economical passenger-carrying operation.

Two 1250 shaft horsepower gas turbine engines furnish economical power for the SK-9. Each engine drives a 9-foot diameter variable pitch propeller and a large lift fan (mounted in the hull) which produces the cushion of air for the 25-ton craft. The twin propeller arrangement permits excellent directional control during maneuvers, with the two rudders providing yaw control. A unique "puff-port" air bleed system pr6vides a lateral control, while pitch and roll trim can be varied while underway by the operator. The inherent stability of the SK-9 makes it easy to handle and the simple control system enables the rapid training of operators.

 BELL AEROSYSTEMS

BUFFALO, NEW YORK – A textron COMPANY

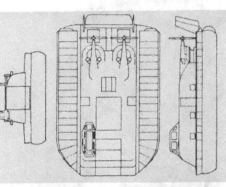

The Bell SK-9B Air Cushion Vehicle is a high performance amphibious craft designed for cargo carrying operations. With a full payload, it will cruise over calm water at 70 MPR, and through 4-to-5 foot waves at 50 MPH. A 90-passenger transport version, the SK-9, is also available.

The SK-9B utilizes many of the proven components of the highly successful S K-5 which has accumulated thousands of hours of operating time throughout the world in commercial and military operations.

The large cargo compartment accommodates 10 tons of low density freight. A loading door is provided in the bow, and the hatch in the roof permits overhead loading. The bow entrance is 7 feet and 6 feet wide, and opens as an integral ramp. The overhead hatch is 7 feet square. In addition, auxiliary doors are provided on each side of the cargo compartment. The multiple entrances provide flexibility to expedite cargo handling.

Two 1250 shaft horsepower gas turbine engines furnish economical power for the SK-9B. Each engine drives a 9-foot diameter variable pitch propeller and a large lift fan (mounted in the hull) which develops the cushion of air for the 25-ton craft. The twin propeller arrangement permits excellent directional control during maneuvers, with the two rudders providing yaw control. A unique "puff-port" air bleed system provides a lateral control, while pitch and roll trim can be varied while underway by the operator. The inherent stability of the SK-9B makes it easy to handle and the simple control system enables the rapid training of operators.

DIMENSIONS:		
Length	55 Feet 8 Inches	
Width	32 Feet 10 Inches	
Height	16 Feet 6 Inches	
Cabin Floor	29 Feet x 17 Feet 8 Inches	
Door Opening	7 Feet 4 Inches x 6 Feet 4 Inches	
	(Plus auxiliary doors)	
WEIGHTS:		
Normal Gross Weight	45,350 pounds	
Overload Gross Weight	52,000 pounds	
Normal Payload	17,700 pounds	
Maximum Payload	23,850 pounds	
Crew	24,000 pounds	
	Two	
POWER PLANT:		
Engine	Two marine gas turbine	
	Maximum Continous Rating 1250 SHP	
Propeller	Two four-blade, variable pitch,	
	9-foot diameter	
Lift Fan	Two 7-foot diameter centrifugal	
Fuel Capacity	660 gallons	
Fuel	Kerosene-type	
PERFORMANCE: (At normal gross weight in I.S.A. conditions)		
Maximum Speed	75 MPH	
Range	225 miles	
Maximum Gradient - Static Conditions	12%	
Cruising Speed in 4-5 Foot Waves	50 MPH	
PERFORMANCE OVER OBSTACLES:		
Solid Wall	3 Feet 6 Inches	
Earth Mound	5 Feet	
Vegetation	6 Feet	
Ditches up to 18 feet wide and 8 feet deep can be crossed at 25 MPH		

4.7   Bell's Air Cushion Landing Craft LCAV

4.7-a   LCAV (Landing Craft, Air Cushion) preparing to move cargo from transport ship to shore

The LCAC is a high-speed ship-to-shore, over-the-beach amphibious air cushion landing craft. Its mission is to move personnel, equipment, and weapons from ships, anchored long distances from the beach, to landing points beyond the beach. Its cargo can include the main battle tank. It is launched, loaded, from the stern of the transport.

243

4.8 Bell's LC JEFF(B) ACV

4.8-a Bell's Data Sheet for LC JEFF(B) ACV

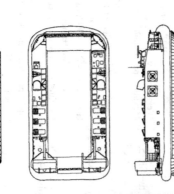

DIMENSIONS:

Length	Overall	80 ft 9in
	Stowed	80 ft
Width	Overall	47 ft
	Stowed	43 ft
Height		23 ft 6in
Cargo area		1738 sq ft
Bow Ramp		28 ft wide
Stern ramp		14 ft 6 in

WEIGHT:

Normal gross weight	325,000 lb
Normal payload	120,000 lb
Overload payload	150,000 lb

CREW:

Four	Operations
Two	Deck

POWER PLANT:

Engine: 6 AVCO Lycoming TF40 Marine Gas Turbines. Normal Continuous Power 16,080 SHP (100F Day)

Propeller: Two four bladed Hamilton Standard 11 ft 9 in Diameter

Fans: Four double-entry centrifugal 5 ft diameter (supplying cushion and bow thrusters)

Fuel Capacity: 6,400 gal

PERFORMANCE:

Speed:	50 knots	in sea state 2 (100F day)
Range:	200 n miles	
Endurance:	3.5 hr at 50 knots	
Max gradient :	13 %	
Wave clearance at 40 knots:		4.5 ft

The LC JEFF (B), a 160-ton air cushion Amphibious Assault Landing Craft (AALC), is an outgrowth of a 10 year U.S. Navy development program. Since 1965 the Navy has been evaluating and defining criteria for an improved amphibious assault landing system in terms of flexibility, speed and cost effectiveness in the movement of material and equipment from ship to shore.

The JEFF (B) will have an operational speed five times that of conventional assault craft and will be able to transition from the sea up onto the beach to designated off-loading points.

The craft will *have* a nominal operating speed of 50 knots and will operate both from the well-decks of landing ships and alongside cargo ships. It has a design payload carrying capacity of 60 short tons with an overload capacity of 75 short tons. The payload will consist of palletized supplies and/or equipment up to the size of the 60-ton Main Battle Tank.

The JEFF (B) is constructed of marine aluminum with propulsion and lift fan systems powered by marine gas turbine engines.

The Naval Sea Systems Command awarded Bell its present contract to design, build and test the prototype in March 1971. The construction phase was completed in early 1977 .

BELL AEROSYSTEMS

A **textron** COMPANY

BUFFALO . NEW YORK

244

4.9 Bell's Model 7501 Viking ACV

4.9-a Bell's Data Sheet for Model 7501 Viking ACV

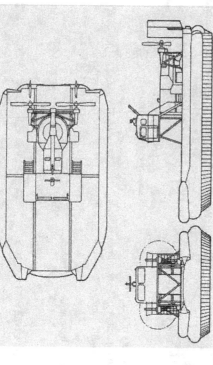

The Viking Air Cushion Vehicle was developed in association with the Department of Industry, Trade and Commerce of the government of Canada.

Viking is fully amphibious, offering economical year round mobility over shallow and deep water, land, ice and snow, and over marginal terrain such as beaches, marshland, sand and mud banks, tundra and scrub.

The distinguished parentage of Viking includes operationally proven ACV's such as the SK-5, SR.N6 and Voyageur. A number of new concepts have been added, including the advanced design tapered skirt, a more efficient puff port system and a V-drive transmission.

The Viking twin drive system reduces propeller speed and noise while maintaining a powerful differential thrust capability for precise controllability.

The versatile flat deck configuration will accommodate a wide range of payloads, superstructures or special equipment providing a unique multi-mission capability applicable to both commercial and military requirements.

Voyageur/Viking component commonality features will simplify maintenance and spares holdings. Viking retains the modular construction features of Voyageur and can be disassembled into easily handled units for deployment by road, rail, sea or air and quickly reassembled oil site,

Bell Aerospace Canada
A DIVISION OF TEXTRON CANADA, LTD.

GENERAL

Operating Crew -- 2 (Commander/Operator, Navigator/Relief Operator)

Maximum cushion pressure -- 39.6 psf

DIMENSIONS

Length (overall) -- 44.5 feet
Beam (overall) -- 26.0 feet
Payload -- Up to 11,000 lb
Height (overall) -- 20.0 feet

WEIGHTS

Weight empty -- 20,685 lb
Maximum permissible gross weight -- 32,500 lb

ROTATING MACHINERY

Power plants -- One UACL ST6T-75 Twin-Pac marine gas turbine (1300 shp continuous; 1700 intermittent, per unit)
Gear Boxes -- Bell/SPECO and SPAR
Propellers -- Two Hamilton Standard three-blade variable pitch, 9 feet diameter
Lift fans -- Two Bell/BHC centrifugal 7 ft. diameter

FUEL SYSTEM

Fuel type -- Kerosene; JP4, JP5, JETA or AVTUR
Capacity -- 1837 U.S. Gals (Equipped for pressure or gravity refueling)

PERFORMANCE (Standard Day and Design Gross Weight)

Maximum calm water speed -- 57 mph
Continuous gradient capability (standard start) -- 10%
Vertical obstacle clearance -- 4 ft
Ditch crossing (width) -- 7 ft
Endurance with Maximum fuel -- 13 hours
Operable in waves up to at least 6 ft
Operable in winds up to 50 mph
Maximum range -- 680 n.m.

CARGO DECK

Area -- 820 sq.ft.
Height (off cushion) -- 3.9 ft.
Tie downs -- 4 rows of 8 ft.
Individual tie down capacity -- 10,000 inner; 5,000 outer

4.10 Bell's LACV-30 ACV

4.10-a Bell's Data Sheet for LACV-30 ACV (Lighter, Amphibian Air Cushion Vehicle – 30 Ton Payload)

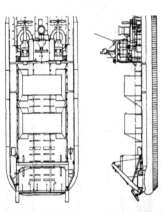

DIMENSIONS:

Length Overall	76 ft 6 in
Width Overall	36 ft 9in
Height Overall	24 ft 8in
Cargo area	51 x 32 ft

WEIGHT:

Design Gross Weight	115,000 lb
Operating Gross Weight	118,000 lb
Maximum Gross Weight	125,000 lb

POWER PLANT:

Engine: Two Pratt & Whitney of Canada Twin-Pac ST6T gas turbines (Max rating 1800 SHP – normal rating 1400 SHP per unit)

Propeller: Two Hamilton Standard Model 43D50 three-bladed variable pitch, 9 ft dia 11 ft 9 in Diameter

Lift Fan: Two Bell/BHC 7 ft centrifugal, 12 bladed, fixed pitch

PERFORMANCE:

Max Payload:	25-30 tons
Cruise Speed:	46 mph
Maximum Speed:	62 mph
Endurance:	8.5 hr

FUEL SYSTEM:

Type: Standard aviation kerosene (Jet A-1,JP4, JP5, JP8, or light diesel oil)

Main Fuel Usable Capacity:	2,240 gal
Fuel Ballast/Emergency Fuel:	1,530 gal
Cruise Fuel Consumption:	260 gal/hr avg
Typical Mission Consumption:	150 gal/hr

ELECTRICAL SYSTEM:

Batteries: Two nickel cadmium, 28 vdc, 40 amp hr

AC Generator: 400 Hz, 115 v

BELL AEROSYSTEMS

BUFFALO, NEW YORK   A Textron COMPANY

LACV-30 (Lighter, Amphibian Air Cushion Vehicle – 30 ton payload) --- militarized Bell "Voyageur" Model 7380.

A fully amphibious, high speed cargo carrier in quantity production at Bell Aerospace Textron, for the U.S. Army.

In 1975, Bell built two prototype craft as a joint development of the company's Niagara Frontier Operations (located near Niagara Falls, N.Y.) and Bell Aerospace Canada Textron in Grand Bend, Ontario. Delivery of 12 craft, forming the U.S. Army 331st Transportation Company at Fort Story, VA was completed in 1983. Twelve additional LACV-30's are now in production with completion scheduled for mid-1986.

Its prime mission for the U.S. Army is as a lighter to move cargo from ship to shore and inland rapidly and efficiently, when port facilities are unavailable. The LACV-30:

• Travels at speeds up to 62 miles per hour, and operates over water, land, snow, ice, even marshes, swamps and low brush, through an 8-foot plunging surf and over 4-foot obstacles.

• Hauls a wide variety of containerized cargo, wheeled and tracked vehicles, engineer equipment, pallets, barrels and other general cargo.

• Can be used for coastal, harbor and inland waterway patrol, search and rescue, medical evacuation, water and fuel resupply; vehicle, personnel and troop transport; augmentation of fixed ports, and pollution and fire control.

• Operates independently of tides, reefs, mud flats, water depth, underwater obstacles or bottom gradients.

• Can be used on 70 percent of the world's beaches compared to the 17 percent now accessible with conventional lighterage craft.

• Permits dry landings for many different payloads.

•An effective icebreaker because of the unique characteristics of its cushion of air.

•Operates effectively even in extremely harsh environments from sub-zero arctic to tropical conditions, including sand beaches and salt water.

• Has demonstrated the best productivity per craft of all the U.S. Army lighterage systems.

• Can be carried fully assembled, as deck cargo on containerships and conventional cargo ships, launched by the ships' crews and readied for service in minutes.

• Can be quickly disassembled into 15 sections, including structural and power modules, side decks, landing pads, flexible skirts and cabin for transport by truck, rail or aircraft.

• Needs no dock or berthing facilities.

• Optional bow-mounted swing crane for self-unloading.

4.11 Bell's SES-100B Surface Effect Ship

4.11-a   Bell's Data Sheet for 100-ton Surface Effect Ship SES-100B

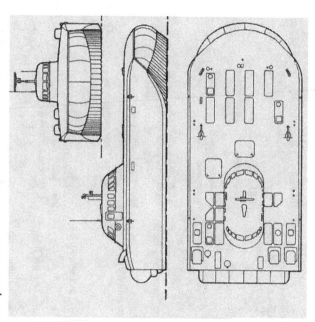

A 100-ton Surface Effect Ship {SES} has been designed and built for the U.S. Navy by the Bell Aerospace Division of Textron, U.S. pioneer in air cushion vehicle design, production and operation since 1958. A radical departure from standard shipbuilding design, materials and techniques makes the craft capable of speeds of 80 knots or more.

The SES-100B rides on a drag reducing cushion of air contained by catamaran-style side hulls and flexible bow and stern seals. When cruising, the center portion of the hull is clear of the water and entirely supported by the air cushion. The air cushion is generated and maintained by a system of eight lift fans. Three marine gas turbine engines provide propulsion by driving two super cavitating propellers located astern of each of the two side hulls.

The test craft was developed under a contract from the U.S. Navy Surface Effect Ships Project Office {PMS-304} and has furnished performance data to confirm the feasibility of constructing and testing a large SES prototype of 3000 tons displacement. The SE8-10oB has successfully operated in a wide range of sea states and has reached a top speed of B9.6 knots (103 mph). It has exhibited maximum seaworthiness and excellent maneuverability throughout its exhausting test program.

Successful development and test of the prototype will lead to Navy acquisition of very high speed, multi-thousand ton ships for a variety of missions. This could result in a small but more effective fleet which would revolutionize Naval warfare.

The SES-100B was launched during 1971 by Bell Aerospace New Orleans Operations at the Michoud Assembly Facility, conducted early tests in Louisiana's Lake Pontchartrain nearby and completed its test program from Bell's Test and Training Facility at the Navy Coastal Systems Laboratory at Panama City, Florida.

DIMENSIONS:
Length   77 ft 8.5 in
Width   35 ft
Height   26 ft 11 in
WEIGHT:
Gross weight (normal)   105 tons
Normal Payload   10 tons
POWER PLANTS:
Propulsion   Three P$W FT 12A-6 marine GT
Lift   Three UACL ST6J-70 marine GT
PERFORMANCE (normal gross weight, standard conditions)
Maximum speed   More than 80 knots
PERSONNEL:
Crew   Four
Observers   Six
MATERIALS:
Hull   Marine Aluminum
Seats   Nitrile PVC/Nylon

Bell Aerospace TEXTRON
Division of Textron Inc.

247

4.12 Bell's Minesweeper Hunter Surface Effect Ship

4.12-a   Bell's Minesweeper Hunter surface effect ship with glass reinforced plastic hull to minimize magnetic and acoustic detection by mines. It was thought that the air cushion design would provide protection from mine explosions. Severe shock impingement testing revealed that the hull tended to delaminate when impacted. In the time that was available, the problem could not be solved. The contract was cancelled.

4.13   Bell's Surface Effect Ships

249

4.14 Bell's Proposal for a Military Tank with Cannon

The size and weight of the vehicle was selected so that it could be transported by a CH-53E helicopter or a C-130E cargo plane: four roof attachment points are available for helicopter lifting. The C-130E could transport one vehicle 3,250 n. mi. while a C5-A can carry six vehicles 4,200 n. mi. The forward track angle will permit the vehicle to climb over an obstacle that is more than 2.5 feet high. Side plates give armor protection to the tracks. A coaxial machine gun is located on the starboard side of the cannon and is intended for internal firing, when the tank is closed up. Multiple grenade launchers are mounted on both sides of the cannon. Fined panels on the port aft side are for engine exhaust and cooling air exhaust.

The maximum forward land speed is 50 mph, on a hard surface. Various types of materials, with slopes greater than 59%, can be climbed without sinking or sliding. Water, up to four feet deep can be safely crossed with the baseline design. Various types of floatation devices could be attached to ensure a swim capability; there would be an associated penalty in vehicle size and weight.

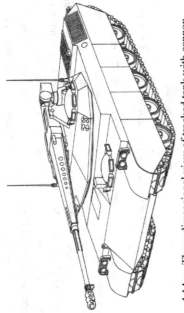

4.14-a Three-dimensional view of tracked tank with cannon.

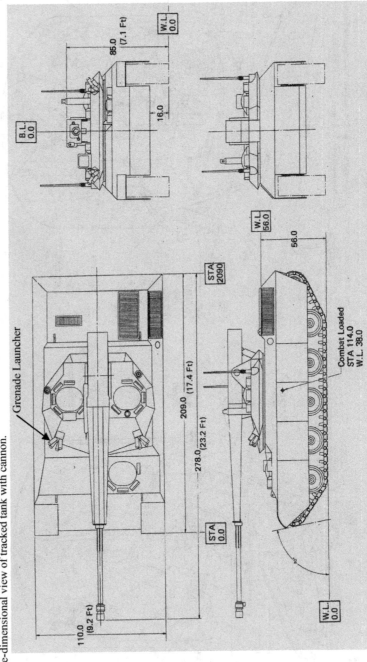

4.14-b External view of baseline tank design with cannon, tracks, two man turret, and forward drivers' station.

250

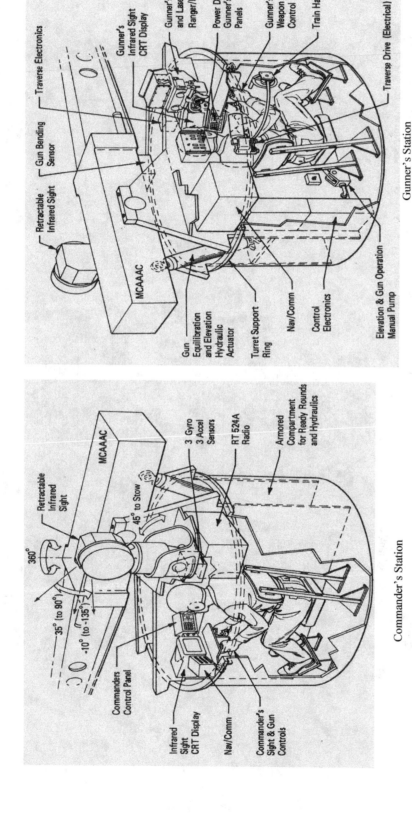

Gunner's Station

Commander's Station

4.14-c Overhead two man 70 inch inside diameter gun turret with a 75mm Ares high velocity cannon that can change elevation from -10° to +35°

The ammunition feed mechanism, of the rapid fire 75 mm cannon, is independent of cannon elevation. This allows for the vertical automated feeding from an armored compartment directly below the cannon. Two different types of ammunition are stored in vertical rows, so the gunner can make a selection of ammunition type; 36 ready rounds of selectable HE and HEAT are available. A blow out panel is installed below the ammunition, to protect tank personnel, in the event the ammunition explodes. A primary day/night FLIR display is installed in the gunner's station, along with a backup dual field-of-view sight with a laser range finder, a control panel and manual backup turret drives for use if the primary electric drive fails. A manually operated hydraulic pump gives backup operation of breech actuation, gun elevation, and feed system actuation.

The commander controls the communication system: a VHF tactical command net radio transceiver, an infantry transceiver, a secure voice modem and the intercom; each crew member has an intercom connection. The commander also has access to the panoramic periscope day/night FLIR CRT display/controls and duplicate computer interface controls for entering commands and numerical data for fire control, and manual range data.

Reference

(1)*Proposal for a Study of Mobile Protected Weapons System New Conceptual Designs*, Bell Aerospace Textron, Report D7619-953001, December 1980, pg. 2-1.

251

4.14-d   Traction system concept 2 – Wheels with run-flat tires. Selection of the type of traction system to be used will be dependent upon final mission objectives.

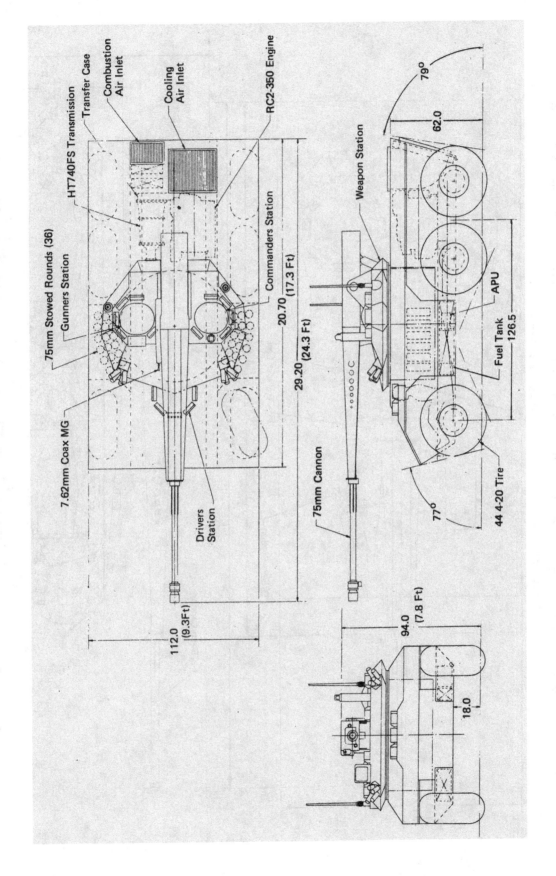

4.14-e   Traction System Concept 3 – Loop Wheel. Selection of the type of traction system to be used will be dependent upon final mission objectives.

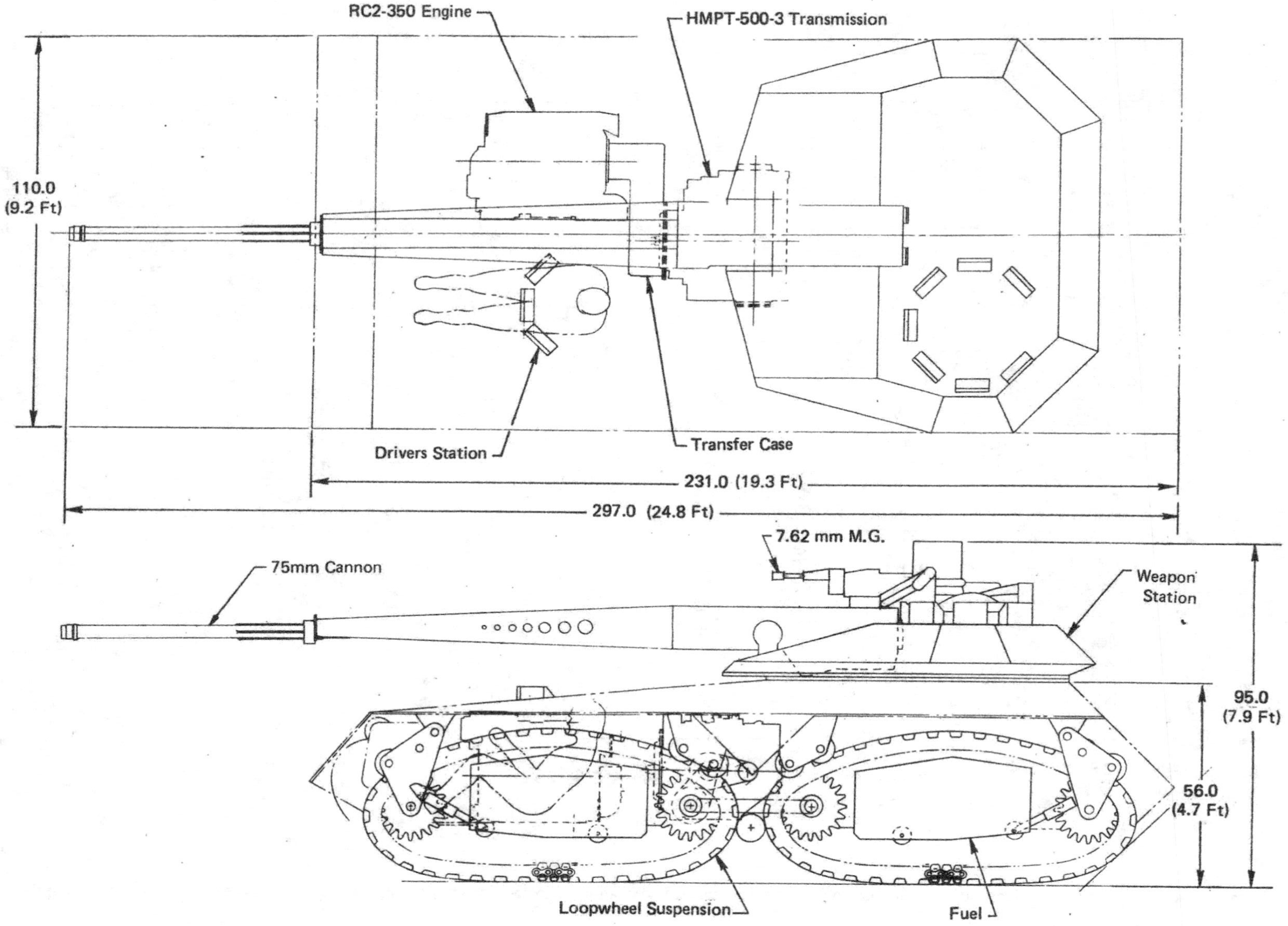

RC2-350 Engine

HMPT-500-3 Transmission

Transfer Case

Drivers Station

231.0 (19.3 Ft)

297.0 (24.8 Ft)

110.0 (9.2 Ft)

Weapon Station

7.62 mm M.G.

75mm Cannon

95.0 (7.9 Ft)

56.0 (4.7 Ft)

Fuel

Loopwheel Suspension

253

4.14-f Bell Aerospace Textron Related Marine Systems Experience

Agency Contract	Program Title	Description	R&D	Analysis	Design	Fabricate	Dev. Test	Op. Eval.	In Service	Marine Engineering Accomplishments	Program Value ($)
USN BuShips NOBS 4501, 4174 4700, 4343	SKMR-1	35 ton ACV	X	X	X	X	X	X		This 35-ton experimental high speed landing craft was designed, constructed and tested by Bell for the Navy. Forerunner of the AALC and LCAC.	6,136,000
USN BuShips/USMC NOBS 4706	ARC(K) X-1	Prototype wheeled amphibious vehicle	X	X	X	X	X			This was a hydrodynamic test bed of a planned, wheeled, high-speed amphibian for the USMC. It was forerunner of the LVA.	140,600
JSESPO DOC MA4678	Surface Effect Ship (SES) Testcraft Fabrication	Detail and fabrication of 100-ton SES testcraft	X	X	X	X	X			Detail design, subcomponent development, construction and test of 100-mph high speed marine testcraft. All performance goals met.	26,383,000
NAVSHIPS N00024-70-C-76	C150-50 Preliminary Design	Conduct preliminary design model tests and full scale mock up of 75 ton payload amphibious landing craft.	X	X	X		X			Preliminary design of lightweight marine structure with comprehensive subsystems to meet exacting Circular of Requirements (COR) from NSRDC. Current successful LC JEFF(B) virtually identical to preliminary design.	1,973,000
NAVSHIPS N00024-71-C-0276	LC JEFF (B)	Design, fabricate and test one 60-ton payload, amphibious landing craft prototype	X	X	X	X	X	X		Hybrid construction of welded and mechanically fastened 100-ton marine aluminum hull. As weighed, completed craft met weight bogey by 3%.	55,100,000
SESPO N00024-73-C-0906	2KSES Preliminary Design	Conduct preliminary design and model tests of 2200-ton SES	X	X	X					Developed detailed plans for design, construction and test of prototype 2KSES.	2,952,000
SESPO N00024-74-C-0909	2KSES Preliminary Design	Conduct detail design, fabrication and test of critical items of 2200-ton SES.	X	X	X	X	X			Development of critical items leading to design and fabrication of the Navy's 2200-ton SES.	36,000,000
U.S. Army (USAMERDC) DAAK02-75-C-0149	LACV-30 Fabrication	Fabrication, spares GSE and Training program for delivery of 2 pilot model ACVs (30-ton payload).	X	X	X	X	X	X		Delivery of pilot model ACVs for U.S. Army logistic application.	4,893,568
U.S. Army (USAMERDC) DAAK07-C-0226	LACV-30 Production	Production quantity of LACV-30 including documentation, logistic, support and depot operation.			X	X	X	X	X	First operational ACV in U.S. military service.	40,000,000
USN NAVSEA N00024-80-C-2284	LCAC Development	LCAC system specification and production plan	X	X	X	X	X	X	X	Design of operational version of Bell-conceived JEFF(B).	3,469,000
Misc. Commercial	BH 110 SES	Proprietary development and production of 110-ft high speed SES-type/crew ferry boats.	X	X	X	X	X	X	X	Incorporation of USN SES technology into moderate-cost, high-performance, reliable commercial craft.	16,100,000
USN NAVSEA N00024-76-C-2067 N00024-77-C-0704 N00024-78-C-2327 N00024-78-C-2562	LVA Phase I Phase II Phase IIb Phase III	Conduct engineering studies, preliminary designs, full-scale mock-up of an air-cushion assisted amphibious armored vehicle.	X	X	X	X	X			Developed credible tracked vehicle design using retractable air-cushion system to provide high overwater transit speed.	1,193,803
NSRDC N00167-80-C-0006	LVT (X) Conceptual Design Study	Develop candidate designs for vehicle to replace the LVTP-7 with emphasis on affordability and superior fighting capability.	X	X	X					Developed candidate designs which met ROC and other requirements.	99,794

254

4.15 Bell's LVA and LVT(X) Amphibious Armored Vehicles

A design study was conducted for the USMC on amphibious armored vehicles that utilized a retractable air cushion system that would provide high speeds over water, employing risk management techniques. A full scale mockup was built. Plans were drawn up for Research, Development, Test, and Evaluation; included were production unit costs and life cycle costs.

4.15-a   LVA amphibious vehicle

A design study was conducted for an amphibious armored vehicle including design trade-offs between armor, armament performance, interior volume, propulsion, and exterior silhouette. As with the LVA design study, risk management techniques were employed and Research, Development, Test and Evaluation plans were drawn up, along with production unit costs and life cycle costs.

Reference

(1) *Proposal for a Study of Mobile Protected Weapons System New Conceptual Designs*, Bell Aerospace Textron, Report D7619-953001, December 1980, pg. 2-1.

255

4.15-b   LVT(X)   amphibious vehicle

4.16 Bell's Manufacturing Proposal for the Hydracobra Light Armored Vehicle

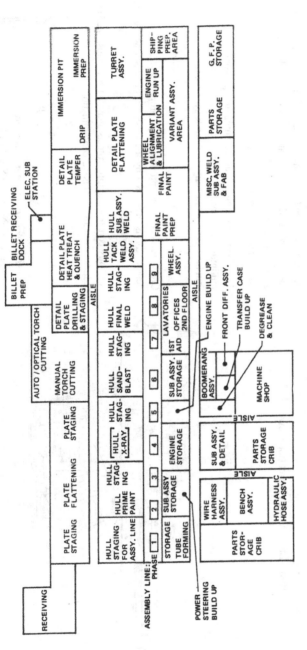

4.16-a Bell's Wheatfield Plant manufacturing floor plan for Hydracobra Program

Bell Aerospace TEXTRON teamed up with Engesa and SAMM to manufacture light armored vehicles for the Marine Corps Rapid Deployment Force. Engesa is an experienced manufacturer of armored cars, 90-mm turrets and guns. SAMM is an experienced manufacturer of 20-30 mm turrets and fire control systems.

For the first year, the light assault turret for the 25-mm gun will be produced by SAMM in France.

Initially, the hull, designated automotive parts, the AG turret and the 90-mm gun will be supplied by Engesa.

Bell's Wheatfield Plane will provide 125,000 square feet to the integration and assembly of the Hydracobra; Figure 4.16-a.

The manufacture of major components will be shifted to American licenses, during the first year, with the goal of complete manufacturing in America.

Variants

A number of variants (Figure 4.16-f) have already been distributed around the world: a field ambulance, a maintenance and recovery vehicle, a motor carrier, and an air defense vehicle. Bell's engineers have been working on upgrading the basic Hydracobra to expand its' capabilities; a variant that expedites the clearing of mine fields, a variant that can self-load two conventional pallets using an integrated winch, and an anti-tank variant.

Firepower

The Light Assault Hydracobra variant has a 20-mm NATA M693 gun installed in a SAMM two-man turret, which provides 360° closed hatch visibility. Standard Hispano-Suisse ammunition is fired from the M693, at a rate of 750 rds/min.

A 90-mm Engresa/Cockerill gun is installed on the Assault Gun variant; anti-material, anti-armor, smoke and canister ammunition are included along with a laser range finder.

256

The 25-mm gun has been stabilized in the Light Assault variant, thereby significantly increasing accuracy while the vehicle is moving.

All three of these variants include a coaxial 7.62-mm machine gun, that is mounted next to the cannon.

Transporting the Hydracobra

The CH-53E helicopter, without any special equipment, can lift and transport the Hydracobra.

Cargo planes can also be used for transport: the C-130 can transport 1 Hydracobra, while the C-141B can transport 3 and the C-5A 7.

Mobility

A diesel engine and an automatic transmission supply a 21.4 hp/ton ratio and an acceleration from 0 to 20 mph in 6.5 seconds, with a top speed of 61mph. Performance has been reliability demonstrated on 30% side slopes, 60% grades, and while crossing 23-inch obstacles. The vehicle has an independent front suspension and a unique walking beam rear suspension.

The Hydracobra swims at 6.2 mph, using propellers and rudders, and maintains a minimum freeboard of 12 inches. Ten feet high surf has been negotiated. The 90-mm gun has been successfully fired while swimming.

Crew Protection

Crew protection is provided by proprietary dual hardness steel armor plates. Penetration is prevented by the hardened outer surface and spalling is prevented by a softer inner layer.

A tightly constructed hull protects against fire bombs, such as Molotov cocktails, and reduces the dangers of NBC contamination.

Reference

(1) *Hydracobra: Light Armored Vehicle Program*, Bell Aerospace TEXTRON/Engesa/SAMM, Dec 1966.

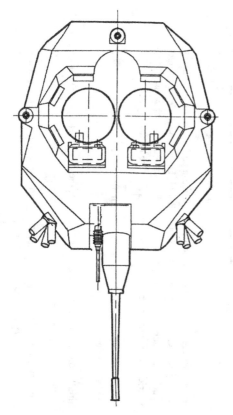

4.16-b   Light Assault Turret (top view)

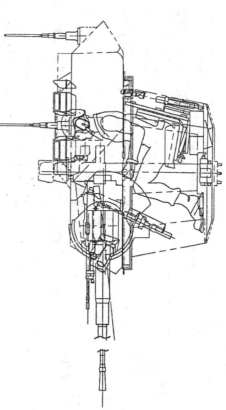

4.16-c   Light Assault Turret (inboard profile)

Characteristics

GENERAL

DRIVE CONFIGURATION - WHEELED	6 × 6
COMBAT WEIGHT - MAX	14.5 TONS
CURB WEIGHT	12.5 TONS
HULL	MONOCOQUE
LENGTH	236 INCHES
WIDTH	102 INCHES
HEIGHT	106 INCHES
WHEELBASE	120.2 INCHES
GROUND CLEARANCE	15 INCHES
ARMOR	DUAL HARDNESS STEEL
CORROSION RESISTANCE	PROTECTIVE FINISHES, PRESSURIZED GEARBOXES
BRAKES	POWER ASSISTED DISC
EMBARKATION/DEBARKATION	2 DOORS: SIDE AND REAR 4 TOP HATCHES DRIVE HATCH
ENGINE	DETROIT DIESEL 6V53T
PRESENTED AREA	44 FT^2 FRONT 80 FT^2 SIDE
TOWING PINTLE	YES
FUEL	DF2

LAND PERFORMANCE

MAXIMUM SPEED	61 MPH
CRUISING RANGE	470 MILES AT 35 MPH
OBSTACLE NEGOTIATION	24 IN. VERTICAL 3.3 FT. TRENCH WIDTH 60° APPROACH ANGLE 42° DEPARTURE ANGLE 9 IN. DIA TREE
SLOPE	67% CLIMB 30% SIDE
ACCELERATION 0 - 20 MPH	7 SECONDS
GRADABILITY	60% SLOPE AT 3.1 MPH
BRAKING FROM 20 MPH	13.5 FT
TURNING RADIUS	30 FT

OVER WATER PERFORMANCE

PREPARATION	NONE
MAXIMUM SPEED	6.2 MPH
RANGE	40 MILES
SURF	10-FT PLUNGING SURF
FREEBOARD	12 IN. AT DRIVERS STATION
CURRENT NEGOTIATION	2.5 MPH
EGRESS/INGRESS	20% SLOPE

ASSAULT GUN

4.16-d   90 mm assault gun variant

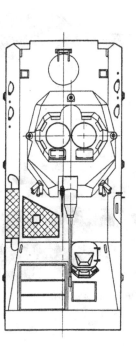

4.16-e   Light assault variant, two man turret, transports six riflemen

Hydracobra

LIGHT ARMORED VEHICLE

Bell Aerospace TEXTRON
Division of Textron Inc.

90-MM ASSAULT GUN

20-MM TURRET-MOUNTED GUN

AMPHIBIAN

TROOP CARRIER

AIR DEFENSE MODULE

ATGM CARRIER

MAINTENANCE/RECOVERY VEHICLE

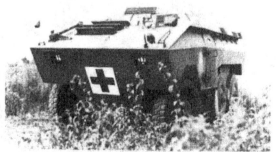

AMBULANCE

4.16-f   Variants of Bell's Hydracobra Light Armored Vehicle

259

5.0   Technology Transfer

5.1 Search for Oil

5.1.1 Bell's Borehole Gravity Meter (BHGM-1)

5.1.1-a Bell's Data Sheet for the BHGM-1 Borehole Gravity Meter, which is used for oil exploration. The instrument is lowered down a pre-bored hole to measure slight variations in gravity, to identify the possible presence of oil.

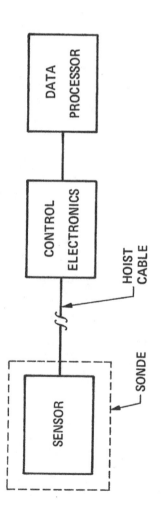

The Bell BHGM-1 borehole gravimeter utilizes the most recent development in inertial guidance technology, the Bell Model XI pendulous accelerometer, to measure gravity variations as low as 5 micro gals.

The BHGM-1 is a self contained instrumentation package containing self leveling, power conditioning temperature control and electrical interface electronics subsystems capable of operating over a wide range of environmental conditions.

Coupled with Bell designed data handling and data reduction system, the BHGM-1 provides real time corrected bore hole gravity data.

Bell received a contract in late 1976 to develop the BHGM for exploring for gas and oil in abandoned oil fields. The intent was to lower the instrument into a dry hole to determine if there still might be oil there. This re-evaluation would be low cost, since the holes would already exist.

The instrument can detect the type of rock formation some distance away. Solid rock has more gravitational attraction than porous rock, which may contain oil.

1. Leveling Performance
   Range $\pm15°$
   Nominal Level ± 10 sec
   Stability ± 10 sec

2. Sensor Accuracy
   Range $\pm 10g$
   Static Reading Accuracy - 5 microgal   rms

3. Operating Environment
   Temperature $-40°F$ to $+200°F$
   Vibration $\pm 1g$
   Magnetic   1 gauss
   Humidity   100% rh
   Pressure   5 psi

4. Power
   Voltage   28 vdc ± 3 vdc
   Power   15 watts

5. Size
   Diameter   3.5 inches
   Length   18 inches
   Weight   3 pounds

6. Data Output
   — Processed (Tide, Drift, Terrain and Anomalous Vertical Gradient Corrected)
     — Interval Gravity
     — Bulk Densities
     — Station Gravity
   — Format - 8 1/2 x 11 Page Printed or
     — Cassette Magnetic Tape
     — Compatible with IEEE STD 488-1975

261

5.1.2 Bell's Gravity Meter-2 (BGM-2)

5.1.2-a Bell's Data Sheet for Gravity Meter-2 (BGM-2), for use in exploring for offshore oil

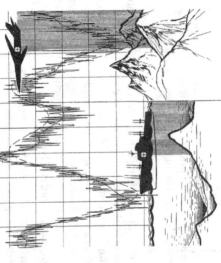

RANGE	966 to 994 gal
ACCURACY	
Static	0.1 mgal
Operational	1.0 mgal
ACCELERATION	
ENVIRONMENT	
(Vertical and Horizontal)	100,000 mgal, periods up to 15 seconds
DRIFT RATE	1.0 mgal/month (predictable)
TARES	None
WEIGHT	445 lb
VOLUME	26 ft3
POWER CONSUMPTION	
Sensor and Stabilization	570 watts
Data Handling	880 watts (typical)
POWER TYPE	28 vdc/or 115V, 60 cps, 1 φ
TEMPERATURE	30° F to 120° F
PRESSURE	30 ±7 in. Hg

The Bell Motion Stabilized Gravity Measuring System, developed by Textron's Bell Aerosystems Company, is designed for precision gravity measurements in a high motion environment on the sea or in the air. One of the primary applications of the system is the rapid acquisition of gravity data which may be utilized in locating high-probability drilling sites for offshore oil wells.

The system, called Bell Gravity Meter 2 (BGM-2), is compact, lightweight, and highly reliable and was developed to military specifications. It consists of a gravity sensor mounted on a two-axis gyro-stabilized platform, and associated electronics; data handling and recording equipment is tailored to the specific requirements of each application. The heart of the gravity sensor is the Bell Model VIIB inertial accelerometer. Over 3,500 Bell accelerometers have been produced for a variety of inertial measurement uses in shipboard, airborne and space applications.

The BGM-2 is an improved version of the BGM-1 system which was delivered to the Gravity Division of the U.S. Naval Oceanographic Office for operational use. Controlled sea trials and operational surveys conducted by the Oceanographic Office have demonstrated the high precision capabilities of the system in a wide variety of sea states. Operational suitability for airborne use was also demonstrated during flight tests conducted in a Navy P3A aircraft.

The BGG-2 incorporates fail-safe monitors and components that are replaceable without adjustment. Self-test circuitry allows the operator to rapidly determine the status of major subsystems, thereby reducing operator monitoring and test time. Designed for use in a wide range of environmental conditions, the modular construction of the system assures high system reliability and effective field maintenance. World-wide system support and service by Bell Aerosystems' personnel is immediately available.

BELL AEROSYSTEMS
BUFFALO, NEW YORK   A textron COMPANY

262

5.2 Ground Communication

5.2.1 Bell's Lightweight Troposcatter Communication Set

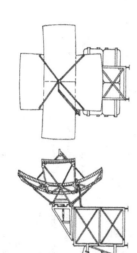

5.2-a Bell's Data Sheet, Lightweight Troposcatter communication set for military ground communications

The AN/TRC-I04 is a lightweight troposcatter communications set designed and built for the Rome Air Development Center by Textron's Bell AerosyHt.ems Company. This lightweight and low power system makes tactical troposcatter communications feasible for military situations in forward areas.

A unique coding diversity transmission system enables the AN/TRC-I04 to operate through a fading medium with relatively small antennas and low transmitter power. Previously, troposcatter communications have been restricted to installations involving large antennas and transmitter powers in the order of kilowatts.

The system consists of two identical terminals each containing a multiplexer, coder, transmitter, receiver and a decoder which provides 12 channels of either voice or digital data transmission and reception. The analog voice signals are multiplexed, sampled, and encoded into five bit digital words. Each five bit digital word is then encoded into a nine digit octal word. Each of the eight symbols in the octal word is transmitted as a distinct frequency, with a two megacycle separation between adjacent frequencies. Because of the redundancy in the code structure the transmitted word is properly decoded if two of the eight frequencies are received.

Since fading is uncorrelated for frequencies separated by two megacycles, there is little probability that more than six frequencies will fade simultaneously. The multifrequency coding scheme provides the AN/TRC-I04 with a time availability of 99.98 percent under the most adverse propagation conditions

RANGE	80 Nautical Miles
TIME AVAILABILITY	99.98 %
NUMBER OF CHANNELS	12
DIGITAL CAPABILITY	Greater Than 2400 Bits/Sec/Channel
ERROR RATE	Less Than 10^{-5}
FREQUENCY BAND	4400 – 4990 MHz
DIVERSITY	8th Order Frequency
MODULATION	Pulse, Multisymbol Error Correcting Code
WEIGHT	490 Pounds
POWER CONSUMPTION	2400 Watts
TRANSMITTER POWER	200 Watts Minimum
ANTENNA GAIN	40 db
MTBF	5120 Hours
MTTR	0.5 Hour
SET UP TIME	41 Minutes

BELL AEROSYSTEMS
BUFFALO, NEW YORK   A textron COMPANY

5.3   Coal Gasification – Converting Coal into Natural Gas

5.3.1   Bell's Development of Coal Gasification Plant

The operation of the system is dictated by the settings of five independent parameters: amount of heat, amount of oxygen, amount of steam, operating pressure, and sometime the need to use a catalyst. The operational set-points are selected according to what type of coal is used and the type of gas that is desired.

An attractive feature of the Bell plant is that it can accommodate all types of coal, including Eastern bituminous coal which is abundant and close by to the Northeast. With 90% of our fossil fuel reserves being coal, coal gasification offers a way for providing natural gas energy for the next 300 years – at the rate of consumption at that time.

Bells' system promises a number of unique features: low capital investment, low operating costs, a build time that is short, a small compact gasifier that is highly efficient and versatile, and a comparatively low cost gas.

Since the start of this project in 1973 funding has been provided by Bell, the Department of Energy, the New York State Energy Research and Development Authority, and the Gas Research Institute.

Bell was working on increasing plant efficiency by upgrading the feed mechanism to create a more uniform flow field. The goal was to continuously consume 12 tons per day of Eastern bituminous coal, mixed with oxygen, for the production of a substitute natural gas or a medium BTU gas.

References

(1) *Bell Aerospace Seeks to Harness Synfuel Energy,* The Buffalo News, February 16, 1980.
(2) Heise, E. H., *Transforming Coal Into Gas and Energy,* National Fuel ENERGY ADVISORY, Vol 3, N0. 2, May 1980.
(3) McCarthy,J., Ferrall, J., Charng, T., Houseman J., *Assessment of Advanced Coal-Gasification Processes,* Jet Propulsion Laboratory, DE82002265, June, 1981.

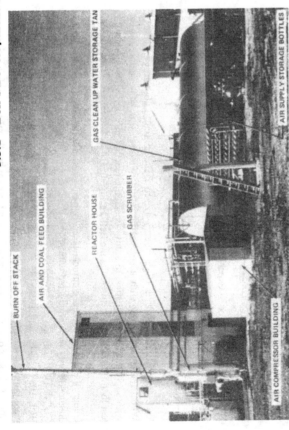

5.3.1-a   Bell coal gasification plant for converting coal into gas

Drawing from its' experience with rocket engine combustion technology, Bell developed a unique process that's employed in its' "High Mass Flux" gasification system. The process relies on a small, compact, highly efficient gasifier which is a cylinder 3.5 feet tall and is capable of converting twelve tons of coal per day.

In operation, pulverized coal, oxygen, and steam are fed into the high pressure and high temperature gasifier. This is referred to as a single stage system, since all the reactants are injected at the same location.

A gas mixture – carbon monoxide, carbon dioxide, hydrogen, and methane -- is produced in the gasifier, along with some sulfur and particulate impurities that are removed by clean-up devices. The end product is an environmentally clean medium BTU gas, which can be used as is or upgraded for use as substitute natural gas. This gas is acceptable for industrial applications or for heating homes.

5.4 Chemical Lasers

5.4.1 Bell's Development of Chemical Laser Technology

A laser is a devise that generates a tight radiant energy beam that can be projected over long distances. Bell worked almost exclusively with infrared lasers; some work was also done with visible lasers. While there are various types of lasers, Bell worked primarily with CW (steady state) infrared chemical lasers, with the intent of significantly increasing the output radiant power. With the goal of eventually developing CW visible chemical lasers, Bell employed a shock tube for preliminary pulsed studies.

Bell's first laser experiment, in 1974, with a CO_2 transfer chemical laser (IRIS) produced more than 15,000 watts; earlier university work produced only 500 watts.

The largest contract, awarded by the US Air Force, was directed at improving the output power per pound of reactants, of a hydrogen fluoride/ deuterium fluoride (HF/DF) multi-purpose chemical laser (MPCL) using facilities at the Wheatfield Plant. This would significantly reduce the weight and volume of a high energy system. Beyond this, Bell investigated the operation of the complete laser system including reactant feed and storage, optics, and pointing the laser and tracking a moving target.

Bell was able to transfer its experience with rocket engines, and including the use of fluorine propellants (Section 3.3), to the chemical laser research. Conceptually, the HF/DF chemical laser is a rocket engine with special mirrors placed at each side of the combustion chamber (Figure 3.1.1-h).

Bell designed and built a High Energy Laser Development Laboratory at the Wheatfield Plant, along with a supersonic nitrogen/helium simulation laboratory, with sophisticated diagnostics, to study supersonic reactant mixing. A unique laser test facility was also constructed at the Bell Test Center.

Under the HF/DF laser contract, laser power, pressure recovery, and various diagnostic tests were conducted with an Air Force mixing nozzle array designated as CL-XI. Bell designed and laser tested three additional nozzles that were designated BCL-10, BCL-16, and BCL-18. Four small Bell nozzles, with different designs, were built for studying the reactant mixing phenomenon, in the laser cavity, under simulated laser operating conditions. For the laser cavity, five water cooled and five non-cooled mirrors were built.

Another contract was awarded to Bell to design and install an MPCL facility at the White Sands Missile Range. This high power laser would be employed for weapons testing and vulnerability testing.

The primary application of this type of laser would probably be the (space based) Star Wars program, where the laser would have to have enough radiant power to destroy an incoming missile or a satellite.

The US Army Ballistic Missile Defense Advanced Technology Center awarded a contract to use the Bell shock tube facility to work on the development of pulsed visible lasers. Earlier, Bell had built their shock tube laboratory and associated diagnostic equipment.

Bell's goal was the eventual production of integrated turn-key operational high energy laser systems.

References

(1) Gross, R.W.E. and Bott, J.F., *Handbook of Chemical Lasers*, John Wiley and Sons, Inc., ISBN 0-471-32804-9, 1976.

(2) Tregay, G. W., Driscoll, R. J., Stricklin, C. E., Cenkner, A. A. Jr., Gribben, E. S., *DF/HF Chemical Laser Technology*, Bell Report No. D9276-927003, Bell Aerospace TEXTRON, Jan 1981.

(3) Cenkner, A. A. Jr., *Cold Flow Visualization and Pitot Probe Scans on a BCL-18 Chemical Laser Trip Nozzle Array*, Bell Repoer No. 9276-928008, Bell Aerospace TEXTRON, May 1980.

(4) Cenkner, A.A. Jr., *Cold Flow Diagnostics on a BCL-10 Chemical Laser Trip Nozzle Array*, Bell Report No. 9276-928005, Bell Aerospace TEXTRON, Oct 1979.

(5) Cenkner, A. A. Jr., *Laser Doppler Velocimeter Diagnostics on a BCL-10 Chemical Laser Trip Nozzle Array*, Bell Report No. 9276-928007, Bell Aerospace TEXTRON, March 1980.

(6) Blauer, J. A., Brandkamp, W.F., Solomon, W.C., *Visible Chemical Laser Development*, Bell Aerospace TEXTRON, May 1979.

(7) Solomon, W.C., *Laser Pioneering: Laser Test Facility at Bell Test Center*, Rendezvous, Bell Aerospace TEXTRON, Vol. XV Spring 1976.

5.4.1-b Exploded view of high energy chemical laser.

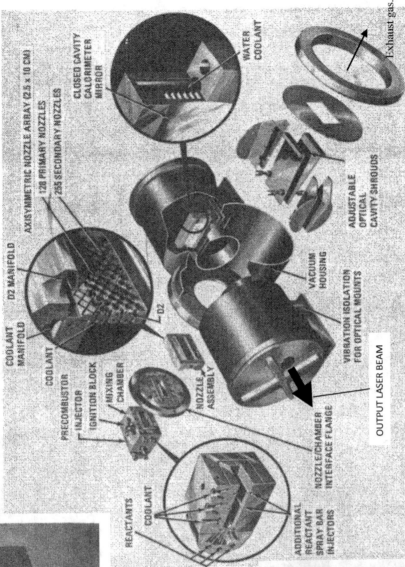

CLOSED CAVITY CALORIMETER MIRROR

WATER COOLANT

AXISYMMETRIC NOZZLE ARRAY (25 x 10 CM)

120 PRIMARY NOZZLES

255 SECONDARY NOZZLES

COOLANT MANIFOLD

D2 MANIFOLD

COOLANT

PRECOMBUSTOR

INJECTOR

IGNITION BLOCK

MIXING CHAMBER

D2

NOZZLE ASSEMBLY

REACTANTS

COOLANT

ADDITIONAL REACTANT SPRAY BAR INJECTORS

NOZZLE/CHAMBER INTERFACE FLANGE

VACUUM HOUSING

VIBRATION ISOLATION FOR OPTICAL MOUNTS

ADJUSTABLE OPTICAL CAVITY SHROUDS

OUTPUT LASER BEAM

Exhaust gas.

Two high speed reactant gas streams enter the laser optical cavity via the nozzle array. In the cavity, they react chemically and emit radiant energy. Mirrors at each end of the laser cavity amplify the emitted radiant energy. Part of the radiant energy is transmitted through one of the mirrors as a usable laser beam. Exhaust gas from the laser cavity is vented to a vacuum, at high speed.

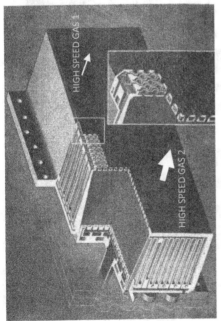

HIGH SPEED GAS 1

HIGH SPEED GAS 2

5.4.1-a Nozzle array.

5.4.1-c Laser test facility at Bell Test Center.

5.4.2   Bell Test Center High Energy Supersonic Chemical Laser

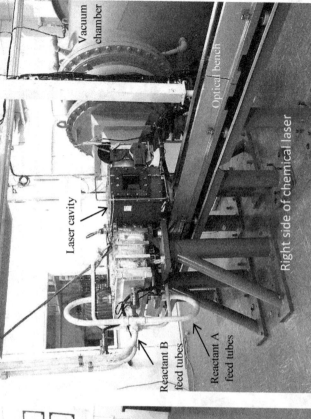

Vacuum chamber

Optical bench

Laser cavity

Reactant B feed tubes

Reactant A feed tubes

Right side of chemical laser

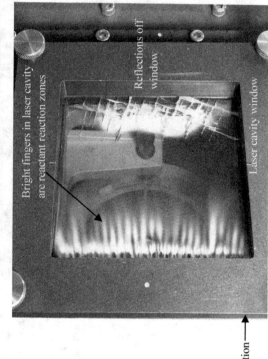

Bright fingers in laser cavity are reactant reaction zones

Reflections off window

Laser cavity window

Close-up of laser cavity during laser operation

267

Probe laser

Pressure gauge

Left side of chemical laser

Diffuser for pressure recovery

Window for diagnostics or for laser power extraction

Laser cavity

Reactant B feed tubes

Reactant A feed tubes

Close-up of supersonic chemical laser

5.4.3 Simulated Reactant Mixing In Supersonic Laser Cavity

5.4.3-a   Diagnostics on Nitrogen/Helium Simulation of Reactants and Gas Trip Jets Entering a Laser Cavity from the Mixing Nozzles

A flow visualization technique was perfected to study interacting regions of a flowing gas mixture. The region of interest was seeded with Iodine gas and then illuminated with a green laser beam. The Iodine would absorb the green light and fluoresce in the yellow. A camera, with the green light filtered out, could then be used to record the regions where the Iodine had spread. This technique proved to be invaluable in determining how gas trip jets enhanced laser power output.   Three interacting nitrogen and helium gas flows simulated the reactants entering, and mixing, in the laser cavity

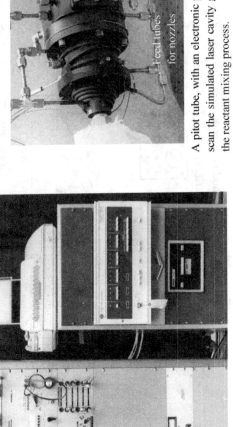

A pitot tube, with an electronic pressure sensor, was developed to automatically scan the simulated laser cavity gas flow field, to help gain an understanding of the reactant mixing process.

268

5.4.4 Shock Tube Laboratory for Simulating Pulsed Laser

5.4.4-a Schematic of shock tube for developing pulsed visible chemical laser

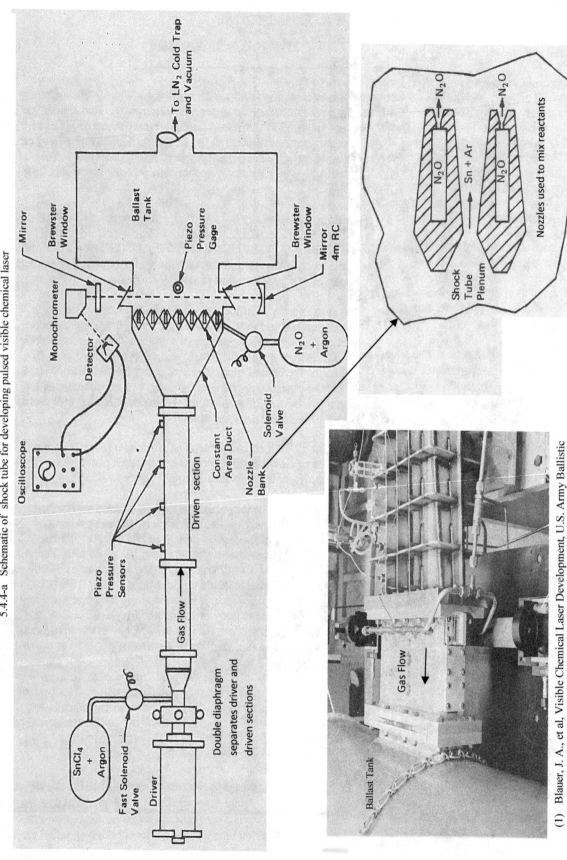

(1) Blauer, J. A., et al, Visible Chemical Laser Development, U.S. Army Ballistic Missile Defense, May 1979.

5.4.5 Multi-Purpose Chemical Laser System (MPCL)

5.4.5-a Multi-Purpose Chemical Laser System to be built by Bell at High Energy Laser Test Facility at White Sands Missile Range, New Mexico. The facility will be used as an automated hydrogen fluoride/deuterium fluoride chemical laser to be used for weapon testing and vulnerability testing.

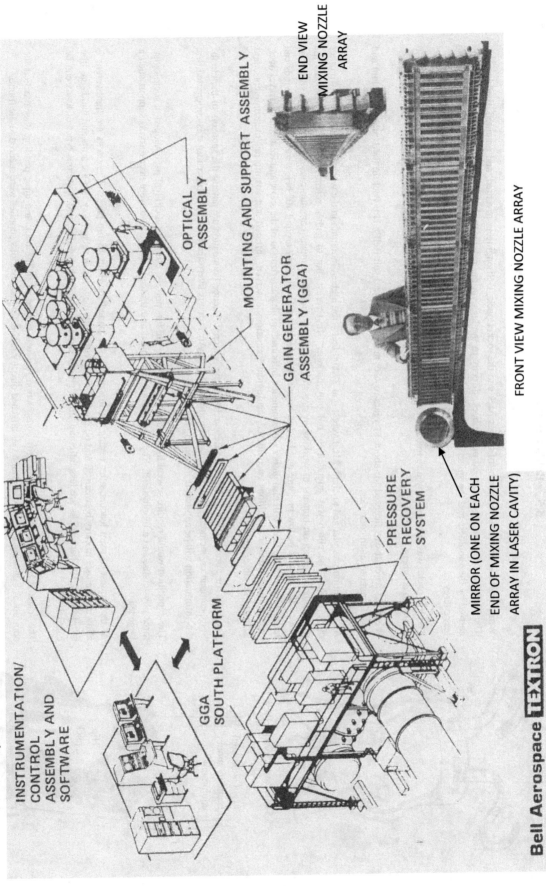

END VIEW
MIXING NOZZLE
ARRAY

OPTICAL
ASSEMBLY

MOUNTING AND SUPPORT ASSEMBLY

GAIN GENERATOR
ASSEMBLY (GGA)

FRONT VIEW MIXING NOZZLE ARRAY

INSTRUMENTATION/
CONTROL
ASSEMBLY AND
SOFTWARE

GGA
SOUTH PLATFORM

PRESSURE
RECOVERY
SYSTEM

MIRROR (ONE ON EACH
END OF MIXING NOZZLE
ARRAY IN LASER CAVITY)

Bell Aerospace TEXTRON

270

5.5 Bell's Hip Pack

While working on the development of the jet pack and the rocket pack, Bell engineers invented, and patented, the hip pack. The hip pack transfers loads, carried on the back, directly to the legs via the pelvis. With traditional back packs the load is carried by the shoulders and the spine. The hip pack allows the user to carry heavier, more balanced, loads while expending less energy.

The hip pack was employed by airborne forces for carrying various types of military equipment. It has also been used by mountain climbers.

Westland Aircraft was licensed for about 14 years to use, manufacture, and sell Bell's Hip Pack in the United Kingdom and most Commonwealth countries.

An expedition was planned to climb Mt. Everest – at 27,890 feet – the highest mountain in the world, in addition to two other close peaks. The climb was intended to, not only conquer Mt. Everest, but to gather scientific information about: the psychological effect on humans of working in stressful situations, and how humans communicate under stressful situations since the air will be low pressure and hard to breath, the temperature will hover around 40° below zero, and the winds can reach 40 mph. The team will also gather environmental information: characterization of the snow and ice that they travel over, high altitude weather conditions encountered, geological formations observed, infrared and cosmic ray measurements, and solar storms observations.

The U.S. Air Force Office of Scientific Research and the National Science Foundation are among the sponsors of the climb. Sixty companies, including Bell, contributed equipment. The Air Force is interested in the climb because it is anticipated that this information will be applicable in training for future space trips.

The expedition would require several hundred men, including guides, porters, climbers, a clinical psychologist, a sociologist, three physicians, and a biophysicist. Eighteen tons of equipment and thirteen tons of food were necessary to support such a large team.

At the request of team leaders, after testing the hip pack, Bell loaned the team 18 hip packs. The packs are lightweight – allowing the user to carry between 50 and 60 pounds -- and can easily be worn in such a way that the arms and hands are free at all times. The back packs are less fatiguing, allow greater mobility and body balance, and numerous types of loads can be carried.

The selected hip pack design was of stiff construction that was shaped to fit the back and hips; padding was attached for comfort. A lightweight aluminum frame was attached directly to the hip pack; various supplies were secured to the frame.

Reference

(1) *Assault on Mt. Everest*, Rendezvous, Bell Aerosystems, Vol. II, No. 1, 1963.

5.5-a Patent for Hip Pack

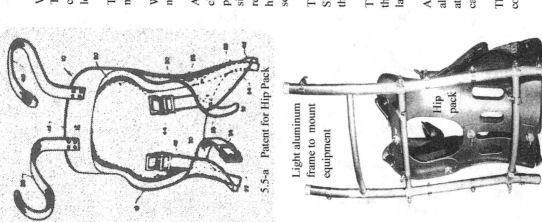

Light aluminum frame to mount equipment

Hip pack

5.5-b Hip Pack Used on Mt. Everest climb

5.6 Identification of Dark Energy

An interdisciplinary approach was taken to identify dark energy – the mysterious, currently unexplained force, discovered in 1998 by professional astronomers, that is responsible for the acceleration of galaxies in the outer reaches of the Universe.

Technology transfer, from experimental and theoretical gas dynamic research in the aerospace industry to the astronomical area, is used to guide the interpretation of Hubble Space Telescope data and leads to classical explanations for numerous unexplained phenomena – in addition to dark energy.

It is shown that dark energy is actually the energy contained in traveling gas cloud shock waves that originate from violent processes that eject gas clouds, like star explosions (e.g. hypernova). The shock wave energy is transferred to a star because, upon impact, a low pressure wake is formed on the downstream side of the star. This leads to an unbalanced force acting on the star, that produces work on the star. This work accelerates the star and eventually leads to the galaxy accelerations reported in 1998.

By studying the pushing behavior of these spherical traveling shock waves, and correlating it with Hubble Space Telescope data and other astronomical data, it is shown that dark energy is also responsible for: revolving star groups and galaxies, Galaxy Streams, Spherical Voids, Walls, Galaxy Clusters, and galaxy collisions. This is because the shock-wave-accelerated-stars are eventually gravitationally captured into orbits, along with other stars, to form galaxies, and other revolving groups. These shock waves also push all the stars out of spherical regions of space, around the explosion site, which creates the Voids.

Experimental work that performed in Bldg. 67 supersonic wind tunnel, at the Bell Wheatfield Plant – as part of the development of high energy lasers – was instrumental in identifying dark energy. The dark energy work was detailed in three published books, which are actually progress reports on the project.

References

(1) Cenkner, August A. Jr., *"Hubble Space Telescope Identifies Dark Energy"*, 3rd edition, ISBN 978-1-4490-1134-5 (sc), 92 pgs, Authorhouse, 8/3/2009.

(2) Cenkner, August A. Jr., *"Dark Energy – Laboratory Simulations Lead to Predictions of: Star Accelerations; Creation of Spiral Galaxies; Formation of Voids, Walls, and Clusters"*, 2nd edition, ISBN 978-1-4343-0661-6 (sc), 194 pgs, AuthorHouse, 8/22/07.

HUBBLE SPACE TELESCOPE
IDENTIFIES DARK ENERGY

August A. Cenkner Jr.
(third edition)

5.7 Bell's Prime Mover

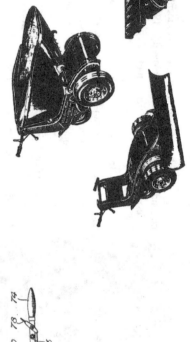

Other models could plow snow and transport loads.

A half ton motorized wheel barrel that could be automatically dumped; the operator rode a dragged plate.

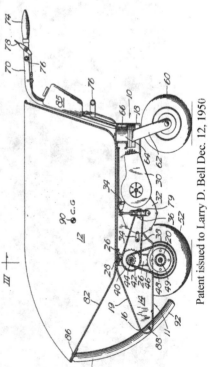

Patent issued to Larry D. Bell Dec. 12, 1950

273

5.8 New Art Form

Computer Generated Modern Abstract Scientific Art*

The art depicts the creation of dark energy -- identified as the kinetic energy contained in a moving spherical gas cloud shock wave – by the death explosion, called a hypernova, of super massive star.

Reasons for Creating the Art

1) Convey scientific information to the viewer.
2) Stimulate discussions of dark energy and science in general.
3) Create some interest in science.

---

* The seed, for this art work, was planted during the research reported in Section 5.4.3. It evolved during the work reported in Section 5.6 and was refined while I wrote this book.

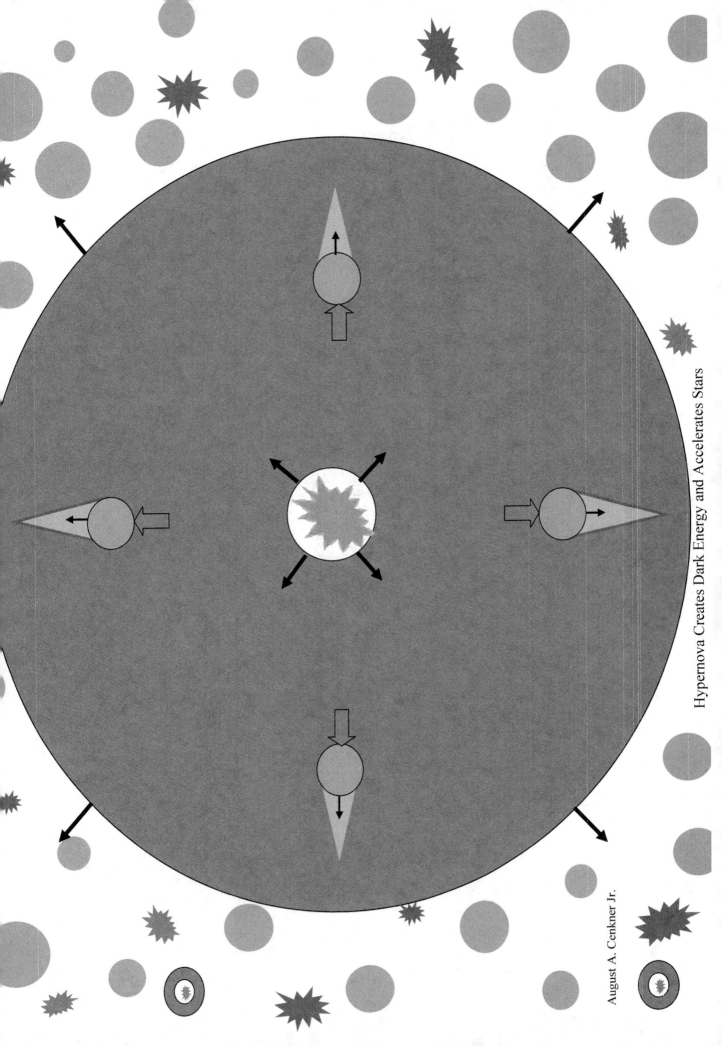

Hypernova Creates Dark Energy and Accelerates Stars

August A. Cenkner Jr.

Description

1.   Remnants of explosion of a massive star (i.e. a hypernova)

2.   Spherical cloud of gas (i.e. a shock wave), ejected from the star during the explosion, moving outward radially at high temperature, high pressure and high speed.

3.   Radial velocity vector showing direction of traveling shock leading edge.

4.   Radial velocity vector showing direction of travel of shock trailing edge.

5.   Four stars trapped inside of spherical traveling gas cloud.

6.   Low pressure wake on back side of star. The high pressure traveling gas cannot expand fast enough to fill in this region.

7.   Velocity vector showing acceleration of star(s) due to the force differential; i.e. high pressure on the front side and low pressure on the back side.

8.   Net accelerating force on star(s) due to pressure differential between high pressure front side and low pressure back side.

9.   Three dimensional field of stars, of different sizes, outside of the spherical cloud of of hypernova gas.

10. Exploding stars (hypernova) in background.

11. Dark energy created by exploding star in background.

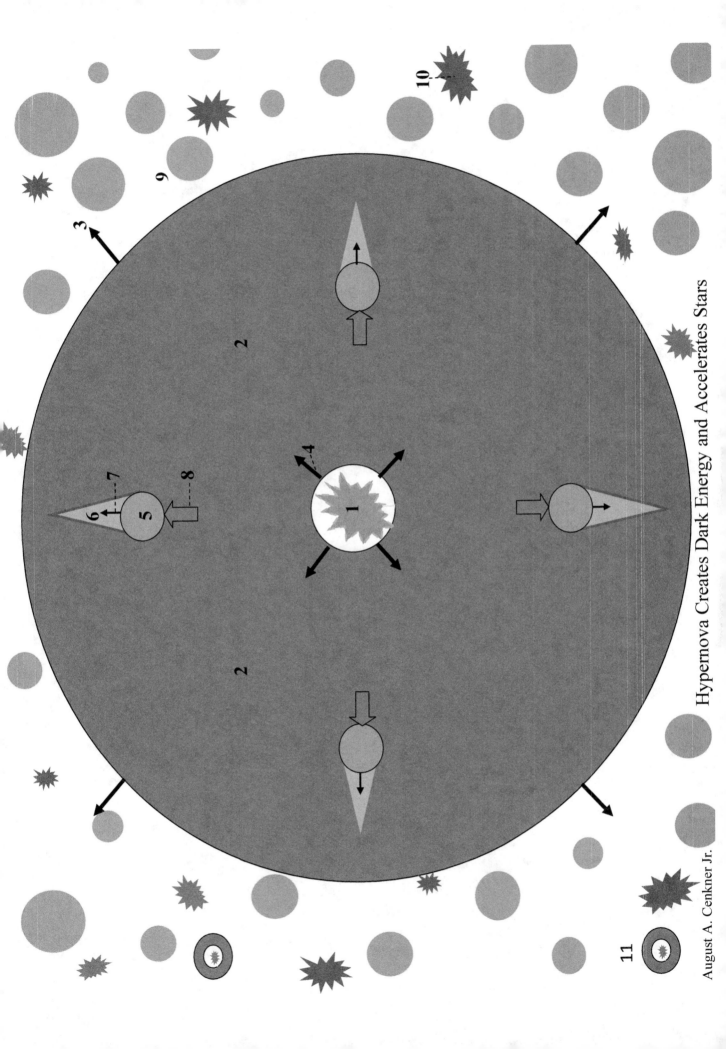

Hypernova Creates Dark Energy and Accelerates Stars

August A. Cenkner Jr.

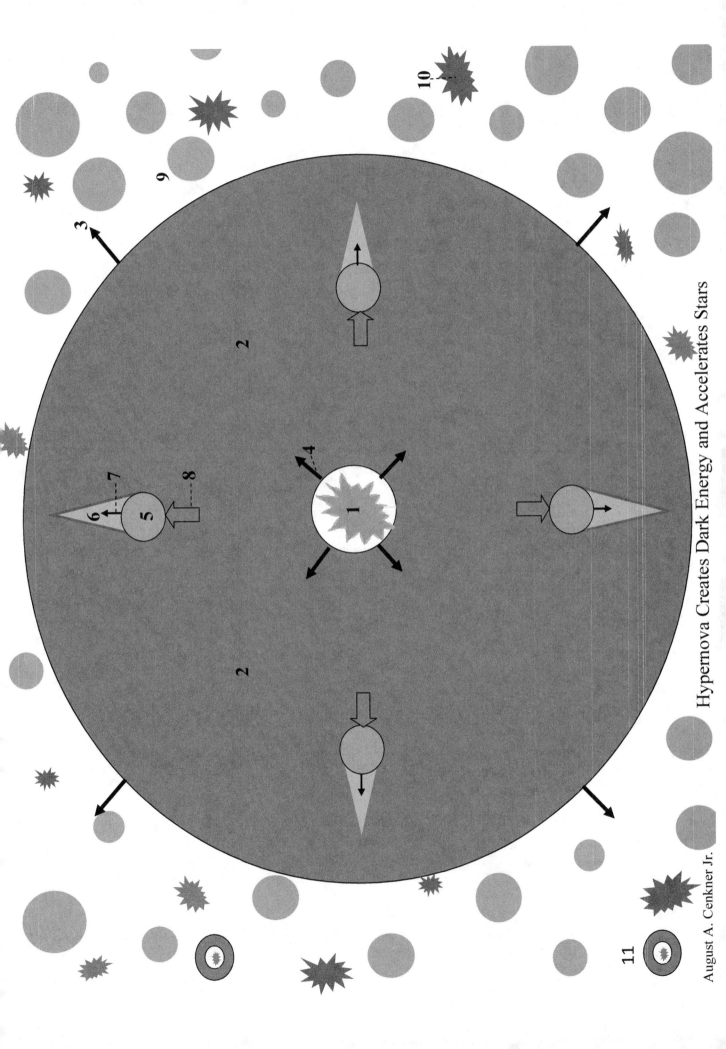

Hypernova Creates Dark Energy and Accelerates Stars

August A. Cenkner Jr.

ABOUT THE AUTHOR

Dr. August A. Jr. and Mrs. Judith B. Cenkner

Dr. Cenkner earned a B.A. degree in Computer Science and a B.S. degree in Aerospace Engineering, along with M.S. and Ph. D. degrees in Engineering Science. He was involved in industrial research and development for forty years, with most of this time being spent in the aerospace industry. He was an ad hoc faculty member, in the School of Engineering at the University of Buffalo, for eleven years -- where he taught thirty-two engineering courses. He worked for Bell Aerospace Textron for five years in the High Energy Laser Department. He was responsible for the operation of the supersonic research facility at Bell, Section 5.4.3, where he conducted extensive research on the simulated mixing of supersonic reactants in the laser cavity. In addition, he conducted laser cavity diagnostics during live firing of high energy lasers.

He recently revisited this laser research and used the results to help develop a theory to explain dark energy. As a result three books, on dark energy, were published; see Section 5.6.